Chest X-ray Made Easy

- Reticulonodular pattern
- Ground glass app / honey comb.
- Kerly B lines.

Commissioning Editor: Laurence Hunter
Project Development Manager: Lynn Watt
Project Manager: Frances Affleck
Designer: Erik Bigland

Chest X-Ray Made Easy

Jonathan Corne MA PhD MB BS MRCP
Consultant Respiratory Physician, Queen's Medical Centre, Nottingham, UK

Mary Carroll MB BS MD MRCP
Consultant Physician, Southampton General Hospital, Southampton, UK

Ivan Brown MB BS FRCS FRCR
Fellow in Radiology , Southampton General Hospital, Southampton, UK

David Delany MA MB Bchir FRCR
Consultant Radiologist, Southampton General Hospital, Southampton, UK

FOREWORD BY
John Moxham MD FRCP
Professor of Respiratory Medicine and Vice Dean,
Guy's, King's and St Thomas' School of Medicine, London, UK

SECOND EDITION

CHURCHILL
LIVINGSTONE

EDINBURGH LONDON NEW YORK OXFORD PHILADELPHIA ST LOUIS SYDNEY TORONTO 2002

CHURCHILL LIVINGSTONE
An imprint of Elsevier Science Limited

Standard edition ISBN 0 443 07008 3

First published 1997
Second edition 2002
   Reprinted 2002, 2003

International Edition ISBN 0 443 07009 1

First published 1997
Second edition 2002
   Reprinted 2002, 2003

British Library Cataloguing in Publication Data
A catalogue record for this book is available from the British Library.

Library of Congress Cataloging in Publication Data
A catalog record for this book is available from the Library of
Congress.

Note
Medical knowledge is constantly changing. As new information
becomes available, changes in treatment, procedures, equipment and
the use of drugs become necessary. The authors and the publishers
have taken care to ensure that the information given in this text is
accurate and up to date. However, readers are strongly advised to
confirm that the information, especially with regard to drug usage,
complies with the latest legislation and standards of practice.

**ELSEVIER
SCIENCE**
your source for books,
journals and multimedia
in the health sciences

www.elsevierhealth.com

The
publisher's
policy is to use
paper manufactured
from sustainable forests

Printed in China
C/03

Foreword

Since the publication of the first edition of this book, nothing has changed my view that the ability of doctors, particularly junior staff, to accurately 'read' the chest X-ray remains crucial. Hospitals have become even busier. Doctors have much less time in which to make decisions about admission, investigation, treatment and discharge of patients. The vast majority of patients admitted to any DGH or teaching hospital have a chest X-ray taken. The clinical decisions affecting the management of these patients are often made before the chest X-ray has been formally reported by radiology departments. Indeed, the chest X-ray is virtually an extension of the physical examination. Faced with a chest X-ray, doctors need to be competent and confident; hence the purpose of this book.

Chest X-ray Made Easy has been successful because the book is clear, concise, focused on the essentials, well illustrated and 'easy' in a constructive way. The second edition builds on the positive qualities of the first, but although this edition improves on its predecessor, the original goal remains the same – to enable a new generation of doctors to excel in the crucial art of interpreting the chest X-ray.

Professor John Moxham,
Professor of Respiratory Medicine,
Vice Dean,
Guy's, King's and St Thomas' School of Medicine,
London

Preface

The chest X-ray is one of the most frequently requested hospital investigations and its intitial interpretation is often left to junior doctors. Although there are a large number of specialist radiology textbooks, very few are targeted at junior doctors and medical students. This book was designed to fill this gap and make interpretation of the chest X-ray as simple as possible. It is not meant as an alternative to a radiological opinion but rather as a guide to making sense of the common abnormalities one is likely to encounter on the wards, for speedy recognition of these will expedite effective treatment of the patient.

Following the success of the first edition we have expanded the book but still kept it small enough to fit in the pocket. Additional sections have been included and abnormalities under the diaphragm are now discussed. The book should remain a useful aid not just for medical students but also for nurses, physiotherapists and radiographers.

Sections 1 and 2 provide some ground rules that must be applied when interpreting the chest X-ray. Section 3 onwards takes the readers through some of the most common abnormalities, arranged according to their X-ray appearance. Each topic contains an example X-ray with an explanatory legend and at the end extra learning points are displayed in the shaded boxes. The outline drawings above the X-rays assist in the interpretation of the abnormality shown.

J. C.
M. C.
I. B.
D.D   2002

Acknowledgements

We would like to acknowledge our colleagues who have read the drafts of this book and made numerous suggestions and contributions, in particular: Kerry Thompson, Fiona Harris, Nicholas Chanarin, Sundeep Salvi, Thirumala Krishna, Peter Hockey, Nicholas Withers, Anoop Chauhan, Bet Mishra, Mark Bulpitt, Sharon Pimento, Anna McKenzie and Vivienne Okaje. We would like to thank Mary Matteson of the Department of Radiology for her work in copying the X-rays and the Department of Teaching Media at Southampton General Hospital for producing the final photographs.

We would also like to thank Professor John Moxham for his invaluable advice with the text and for writing the Foreword, and staff at Harcourt Health Sciences.

Contents

How to look at a chest X-ray

Basic interpretation is easy

Basic interpretation of the chest X-ray is easy. It is simply a black and white film and any abnormalities can be classified into:

1. Too white.

2. Too black.

3. Too large.

4. In the wrong place.

To gain the most information from an X-ray, and avoid inevitable panic when you see an abnormality, adopt the following procedure:

1. Check the name and the date. Do this before you put it on the screen — if you do not you are bound to forget.

2. Check the technical quality of the film. (Explained in Chapter 1.2.)

3. Scan the film thoroughly and mentally list any abnormalities you find. Always complete this stage. The temptation is to stop when you find the first abnormality but if you do this you may get so engrossed in determining what it is that you will forget to look at the rest of the film. Chapter 1.3 expains how to scan a film.

4. When you have found the abnormalities, work out where they are. Decide whether the lesion is in the chest wall, pleura, within the lung or mediastinum. Chapter 2 explains how to localize lesions within the lung and the heart, Chapter 7 the mediastinum, and Chapter 8 the ribs.

5. Mentally describe the abnormality. Which category does it fall into:
   I   Too white.
   II   Too black.
   III Too large.
   IV In the wrong place.

Chapters 3 to 10 will take you through how to interpret your findings.

6. Always ensure that the film is reported on by a radiologist. Basic interpretation of the chest X-ray is easy but more subtle signs require the trained eye of a radiologist. Seeking a radiologist's opinion can often expedite a diagnosis.

7. Finally do not forget the patient. It is possible and indeed quite common for a very sick patient to have a normal chest X-ray.

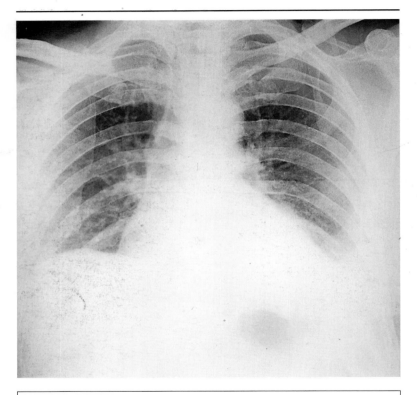

The next four X-rays are examples of how the technical quality of a film can affect its appearance and potentially lead to misinterpretation. Above is an AP film which shows how the scapulae are projected over the thorax and the heart appears large. Compare this to the film opposite which is a standard PA projection showing how the scapulae no longer overlie the thorax and the heart size now appears normal.

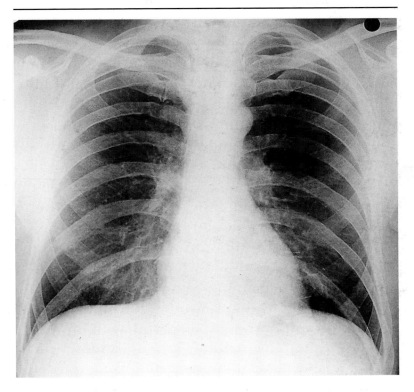

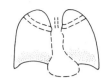

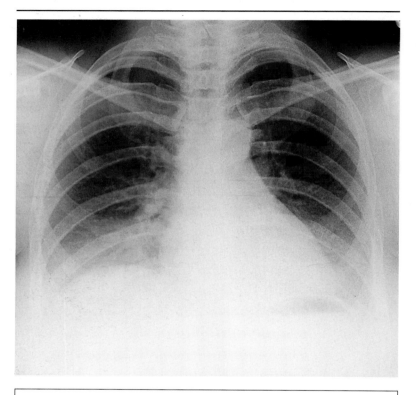

Films on pp 6 and 7 show the effect of respiration. The film above has been taken during a poor inspiratory effort whereas the film opposite has been taken during full inspiration.

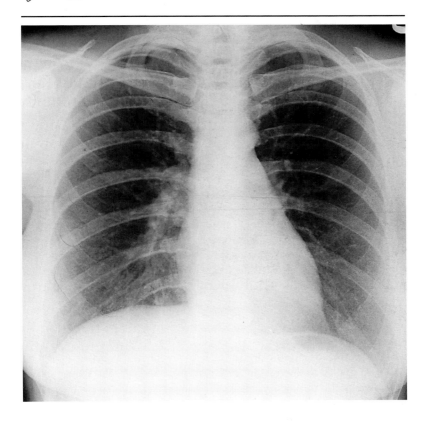

Always check the technical quality of any film before interpreting it further. To do this you need to examine in turn the projection, orientation, rotation, penetration and degree of inspiration. Problems with any of these can make interpretation difficult and unless you check the technical quality carefully you may misinterpret the film.

Projection

Look to see if the film is anteroposterior (AP) or posteroanterior (PA). The projection is defined by the direction of the X-ray beam in relation to the patient. In an AP X-ray the X-ray machine is in front of the patient and the X-ray film at the back. In a PA film the beam is fired from behind the patient and the film placed in front. The standard chest X-ray is PA but many emergency X-rays are AP because these can be taken more easily with the patient in bed. AP films are marked AP by the radiographer and PA films are often not marked since this is the standard projection. If you are not sure then look at the scapulae. If the scapulae overlie the lung fields then the film is AP. If they do not it is probably PA. If the X-ray is AP you need to be cautious about interpretation of the heart size and shape of the mediastinum since both can be distorted. An AP film can be taken with the patient sitting or lying. The film should be marked erect or supine by the radiographer. It is important to note this since the appearance of a supine X-ray can be very different to that of an erect one.

Orientation

Check the left/right markings. Do not assume that the heart is always on the left. Dextrocardia is a possibility but more commonly the mediastinum can be pushed or pulled to the right by lung pathology. Radiographers always safeguard against this by marking the film left and right. Always check these markings when you first look at the film but remember the radiographer can sometimes make mistakes — if there is any doubt re-examine the patient.

Rotation

Identify the medial ends of the clavicles and select one of the vertebral spinous processes that falls between them. The medial ends of the clavicles should be equidistant from the spinous process. If one clavicle is nearer than the other then the patient is rotated and the lung on that side will appear whiter.

Penetration

To check the penetration, look at the lower part of the cardiac shadow. The vertebral bodies should only just be visible through the cardiac shadow at this point. If they are too clearly visible then the film is over penetrated and you may miss low density lesions. If you cannot see them at all then the film is under penetrated and the lung fields will appear falsely white. When comparing X-rays it is important to check that the level of penetration is similar.

Degree of inspiration

To judge the degree of inspiration, count the number of ribs above the diaphragm. The midpoint of the right hemidiaphragm should be between the 5th and 7th ribs anteriorly. The anterior end of the 6th rib should be above the diaphragm as should the posterior end of the 10th rib. If more ribs are visible the patient is hyperinflated. If fewer are visible the patient has not managed a full intake of breath perhaps due to pain, exhaustion or disease. It is important to note this, as a poor inspiration will make the heart look larger, give the appearance of basal shadowing and cause the trachea to appear deviated to the right.

Scanning the PA film

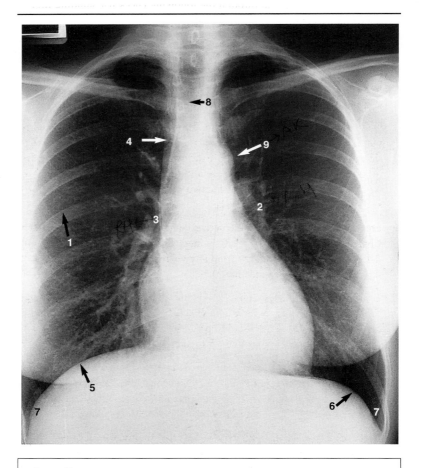

The PA film

Find a decent viewing box with a functioning light that does not flicker. If possible lower the ambient lighting. In order to recognize areas that are too white or too black you need to survey the X-ray from a distance (about 4 ft/ 1.2m) and then repeat this close up. Go through the following check list standing first at about 4 ft/1.2m from the X-ray and then close up.

1. *Lung fields.* These should be of equal transradiancy and one should not be any whiter or darker than the other. Try to identify the horizontal fissure (1) (this may be difficult to see) and check its position. It should run from the hilum to the sixth rib in the axillary line. If it is displaced then this may be a sign of lung collapse.

   One of the first signs of lung disease is loss of volume of that lung and so you need to determine whether either of the lung fields are smaller than they should be. This is difficult since the presence of the heart makes the left lung field smaller. As you see more and more chest X-rays, however, you will gain an appreciation of how the two lung fields should compare in size and therefore be able to detect when one is smaller than it should be.

   Look for any discrete or generalized shadows. These are described in Chapter 3 — The white lung field. Remember that abnormalities in the lung field can represent pathology from anywhere from the cortex of the rib to the outer edge of the mediastinum.

2. *Look at the hilum.* The left hilum (2) should be higher than right (3) though the difference should be less than 1 in/2.5 cm. Compare the shape and density of the hila. They should be concave in shape and look similar to each other. Chapter 5 describes how to interpret hilar abnormalities.

3. *Look at the heart.* Check that the heart is of a normal shape and that the maximum diameter is less than half of the transthoracic diameter. Check that there are no abnormally dense areas of the heart shadow. Section 6 takes you through interpretation of the abnormal heart shadow.

4. *Check the rest of the mediastinum.* The edge of the mediastinum should be clear though some fuzziness is acceptable at the angle between the heart and the diaphragm, the apices and the right hilum. A fuzzy edge to any other parts of the mediastinum suggests a problem with the neighbouring lung (either collapse or consolidation) dealt with in Chapter 3. Interpretation of the widened mediastinum is dealt with in Chapter 7.

Look also at the right side of the trachea. The white edge of the trachea (4) should be less than 2–3 mm wide on an *erect* film. (See Chapter 1.1 for interpretation.)

5. *Look at the diaphragms.* The right diaphragm (5) should be higher than the left (6) and this can be remembered by thinking of the heart pushing the left diaphragm down. The difference should be less than 1.2 in/3 cm. The outline of the diaphragm should be smooth. The highest point of the right diaphragm should be in the middle of the right lung field and the highest point of the left diaphragm slightly more lateral.

6. *Look specifically at the costophrenic angles* (7). They should be well defined acute angles.

7. *Look at the trachea* (8). This should be central but deviates slightly to the right around the aortic knuckle (9). If the trachea has been shifted it suggests a problem within the mediastinum or pathology within one of the lungs.

8. *Look at the bones.* Step closer to the X-ray and look at the ribs, scapulae and vertebrae. Follow the edges of each individual bone to look for fractures. Look for areas of blackness within each bone and compare the density of the bones which should be the same on both sides.

9. *Soft tissues.* Look for any enlargement of soft tissue areas.

10. *Look at the area under the diaphragm.* Look for air under the diaphragm or obviously dilated loops of bowel. Remember that abdominal pathology can occasionally present with chest symptoms.

How to look at the lateral film

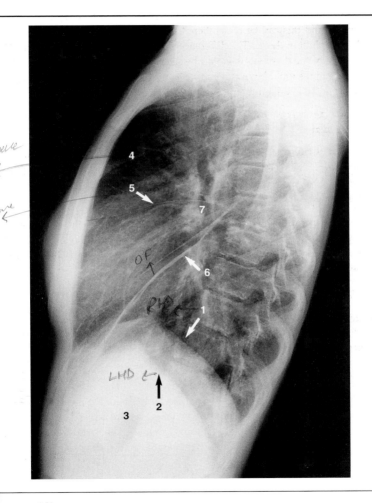

Lateral film

A lateral chest X-ray can be taken with either the right or left side of the patient against the film. Do not worry about which way it has been taken since for all but the most subtle signs it makes little difference. It is useful to get into the habit of always looking at the film the same way and we suggest looking at the film with the vertebral column on the right and the front of the chest on the left. Once you have done this:

1. Check the name and the date.

2. Identify the diaphragms. The right hemidiaphragm (1) can be seen to stretch across the whole thorax and can be clearly seen passing through the heart border. The left (2) seems to disappear when it reaches the posterior border of the heart.

3. Another method of identifying the diaphragms is to look at the gastric air bubble (3). Look again at the PA film and work out the distance between the gastric air bubble (which falls under the left diaphragm) and the top of the left diaphragm. Make a note of this. Now go back to the lateral. The diaphragm that is the same distance above the gastric air bubble is the left diaphragm.

You can now set about interpreting the film. As with the PA step back from the film and adopt the following process:

1. Compare the appearance of the lung fields in front of and above the heart to those behind. They should be of equal density. Check that there are no discrete lesions in either field.

2. Look carefully at the retrosternal space (4). An anterior mass will obliterate this space turning it white.

3. Check the position of the horizontal fissure (5). This is a faint white line which should pass horizontally from the midpoint of the hilum to the anterior chest wall. If the line is not horizontal the fissure is displaced. Check the position of the oblique fissure (6) which should pass obliquely downwards from the T4/T5 vertebrae, through the hilum, ending at the anterior third of the diaphragm.

4. Check the density of the hila (7). A hilar mass may make the hila whiter than usual.

5. Check the appearance of the diaphragms. Occasionally a pleural effusion is more obvious on a lateral film. Its presence would cause a blunting of the costophrenic angle either anteriorly or posteriorly.

6. Look at the vertebral bodies. These should get more translucent (darker) as one moves caudally. Check that they are all the same shape, size and density. Look for collapse of a vertebra or for vertebrae that are significantly lighter or darker than the others.

Localizing lesions

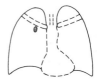

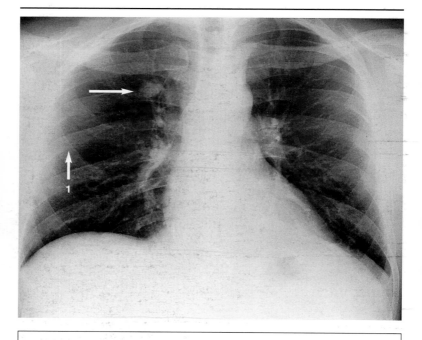

This pair of films shows a nodule in the upper lobe of the right lung (arrow). The PA film shows that it is situated above the level of the horizontal fissure (1) and the lateral film shows that it lies in front of the oblique fissure (1) and above the horizontal fissure (2).

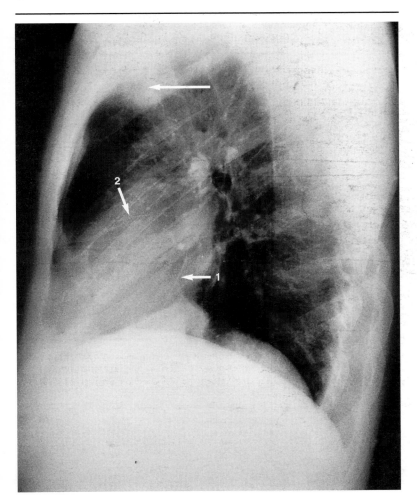

As well as knowing what a lesion is it is often important to know its position within the lung. To accurately localize a lesion on a chest X-ray you need to look at both the PA and lateral films. First look at the PA film:

1. The position of the lesion can be described in terms of zones. The upper zone lies above the right anterior border of the 2nd rib, the middle zone between the right anterior borders of the 2nd and 4th ribs and the lower zone between the right anterior border of the 4th rib and the diaphragm. Although this is useful descriptively it does not give any information about the lobes of the lung.

2. Look at the borders of the lesion. If the lesion is next to a dense (white) structure then the border between the lesion and that structure will be lost — this is called the silhouette sign. Therefore if the lesion is in the right lung and obscures part of the heart border it must be in the right middle lobe. If it obscures the border of the diaphragm it is in the right lower lobe.

Now look at the lateral — you *must* do this to accurately localize the lesion.

If the lesion is in the right lung:

1. Identify the oblique fissure (1) (p. 18). If the lesion lies posterior to the oblique fissure it must lie within the lower lobe no matter how high it appears on the PA film.

2. If the lesion lies anterior to the oblique fissure it may be in the upper or middle lobe. Identify the horizontal fissure (2) (p. 19). If the lesion is below the horizontal fissure it is in the middle lobe. If it is above it is in the upper lobe.

If the lesion is in the left lung:

1. Identify the oblique fissure (p. 19). If it is behind the oblique fissure it must be in the lower lobe. If it is anterior to the oblique fissure it is within the upper lobe — there is no middle lobe on the left!

The heart

In order to fully assess any abnormalities of the shape of the heart it is important to understand the composition of the heart shadow. Look at the following four films on pages 22–25.

1. Look at the right heart border and follow it up from the diaphragm. From the diaphragm to the hilum the heart border is formed by the edge of the right atrium (1). From the hilum upwards it is formed by the superior vena cava (2).

2. Follow the heart left border up from the diaphragm. From the diaphragm up to the left hilum it consists of the left ventricle (3). The left border is then concave at the lower level of the left hilum and here it is made up of the left atrial appendage (4). This concavity is lost when the left atrium is enlarged leading to a straightening of the left heart border and sometimes the development of a convexity at this point. At the level of the hilum the border is made up of the pulmonary artery (5) and above this the aortic knuckle (6).

The lateral film is useful. The posterior border of the heart shadow is made up of the left ventricle (7) and the anterior border the right ventricle (8). To identify whether any lesion is in the mitral or aortic valve draw an imaginary line from the apex of the heart to the hilum. If the abnormal valve lies above this line it is aortic and if it lies below, it is mitral.

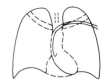

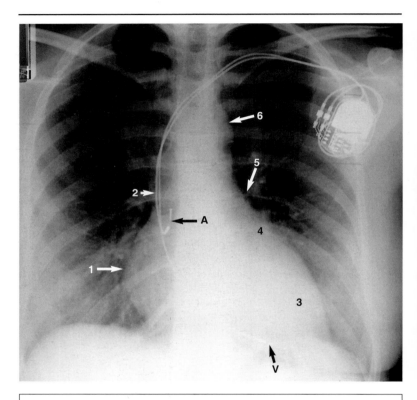

Film 1 is an X-ray of a patient with atrial (A) and ventricular (V) pacing wires and demonstrates the position of the right atrial appendage and ventricle in the PA and lateral films.

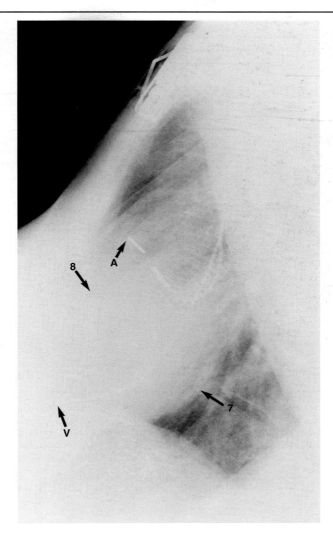

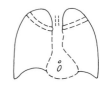

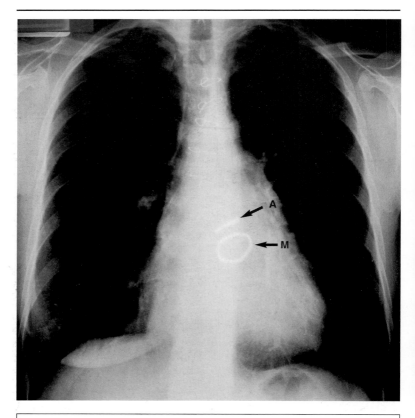

Film 2 is of a patient with prosthetic aortic (A) and mitral (M) valves showing their position in the AP and lateral films.

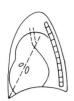

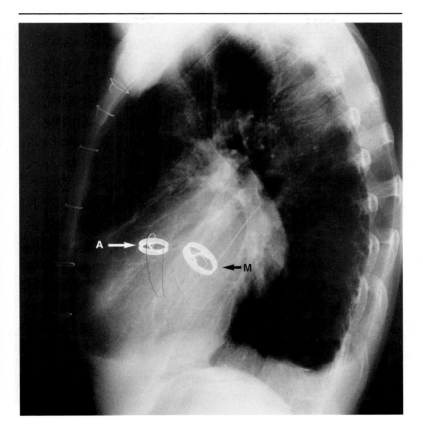

Chapter 3

The white lung field

3.1

Pleural effusion

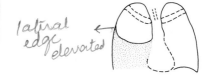

lateral edge elevated

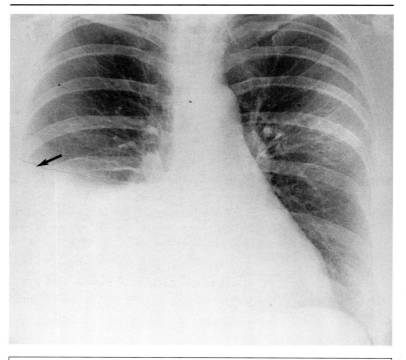

This film shows a moderately large right pleural effusion. The right lung field is white with a meniscal edge (arrowed) suggesting an effusion. The mediastinum is shifted away from the effusion.

consolidation = heterogenous + air bronchogram

If you see an area of whiteness at the base of a lung then the possible causes are a pleural effusion, a raised hemidiaphragm and an area of consolidation or collapse. You need to determine which of these it is.

1. Look closely at the texture of the whiteness. Consolidation usually causes more heterogeneous shadowing typically with the presence of an air bronchogram. Look carefully for an air bronchogram (p. 44) since its presence will point to consolidation rather than a pleural effusion.

2. Look at the shape of the upper border of the shadowing. Fluid will have a meniscus so the upper outer border of an effusion will be concave.

3. To differentiate an effusion from a raised hemidiaphragm look again at the shape of the upper border. The upper border of an effusion will peak much more laterally than you would expect the diaphragm to do. This is a matter of looking at lots of X-rays.

4. Look for mediastinal shift. It can be difficult to differentiate an effusion from lung collapse. Collapse usually causes mediastinal shift towards the white lung field so the absence of shift suggests the presence of an effusion. Remember, however, that collapse can accompany an effusion so that although the absence of shift implies an effusion its presence does not exclude it.

5. A lateral view is often helpful since the meniscus on a lateral can be much more obvious. Look for the presence of a meniscus which, often on the lateral, is seen to tent up into one of the fissures.

6. If you diagnose an effusion look on the X-ray for possible causes. Check the size of the heart (a large heart points to heart failure), look at the hilum for possible enlargement. Look at the visible parts of the lung fields for obvious masses and check the bones for signs of metastasis. Look very carefully at the apices of the lungs for tumours and TB.

Causes of a pleural effusion

Transudate <30 g/l of protein
Heart failure, e.g. congestive cardiac failure, pericardial effusion
Liver failure, e.g. cirrhosis
Protein loss, e.g. nephrotic syndrome, protein losing enteritis
Reduced protein intake, e.g. malnutrition
Iatrogenic, e.g. peritoneal dialysis

Exudate >30 g/l of protein
Infection, e.g. pneumonia, tuberculosis
Infarction
Malignancy, e.g. bronchial carcinoma, mesothelioma, metastasis
Collagen vascular disease, e.g. rheumatoid arthritis, SLE
Abdominal disease, e.g. pancreatitis, subphrenic abscess
Trauma/surgery

Collapse

Collapse of a lung is an important cause of a white lung on X-ray. When confronted with a white lung it is important to be thorough in looking for the features suggestive of collapse since the presence of collapse indicates possible serious pathology.

Collapse of the lung leads to a loss of volume of that part of the lung and so the normal radiological landmarks will be distorted. To diagnose collapse look at each of these markings carefully and decide whether they are in the correct position. You will need to look at the lateral X-ray as well as the PA.

On the PA film:

1. Look at the lung fields. The right lung should be larger than the left — if it is not, suspect an area of right-sided collapse.

2. Look at the diaphragms. The right diaphragm should be higher than the left. Collapse in the left lung may distort this.

3. Look for the horizontal fissure in the right lung (pp 11, 14). The horizontal fissure on the right should run from the centre of the right hilum to the level of the 6th rib at the axillary line. If this is pulled up it suggests right upper lobe collapse or if pulled down right lower lobe collapse.

4. The heart should straddle the midline with one-third to the right and two-thirds to the left. The heart shadow will be deviated to the side of collapse. Look at old films. Mediastinal shift is much easier to detect by comparing with the patient's old, normal X-rays.

5. The heart borders should be distinct. If the lung adjacent to the heart is collapsed then the heart border will appear blurred. If the right heart border is blurred this indicates right middle lobe collapse and if the left is blurred, lingular collapse (see opposite).

6. The trachea should be central. Collapse of the right or left upper lobes will pull the trachea towards the area of collapse. Again this may be easier to spot by comparing with the patient's old films.

On the lateral film:

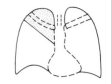

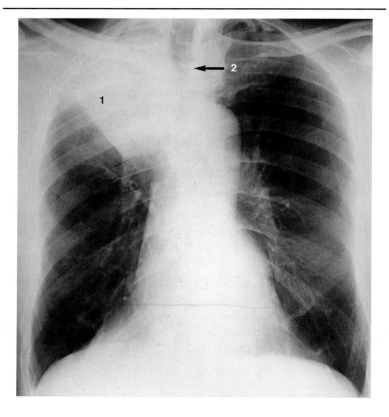

Check the position of the oblique and horizontal fissures (pp 11 and 14). Any displacement from their normal position suggests collapse. Collapse of any of the lobes of the lung gives a distinct appearance on the X-ray. These are described below. (Do not worry about the appearance of lingular collapse. It is extremely rare and you are unlikely to encounter it.)

See also **Volume loss** 'learning point', p. 41.

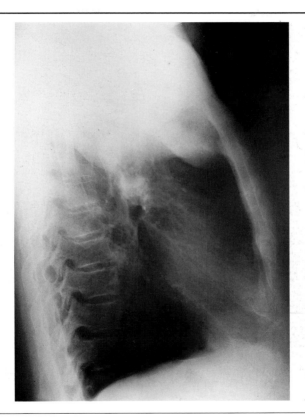

1. **Right upper lobe collapse.** *Facing page: There is an area of whiteness in the upper zone of the right lung (1). The horizontal fissure is elevated, there is an apparent right upper hilar 'mass', the trachea is deviated to the right (2) and the ribs over the area of whiteness are closer together than is normal. On the lateral film (above) the increased whiteness in the uppermost part of the chest may be seen.*

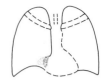

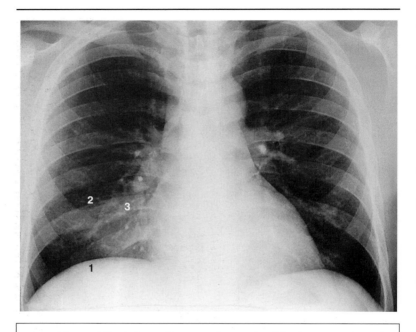

2. **Right middle lobe collapse.** This can be difficult to spot. The right diaphragm may be slightly raised (1) and the horizontal fissure (2) may be lower than usual. The upper part of the lower zone may have a hazy white appearance (3) and the heart border is sometimes indistinct. It is easier to detect in the lateral film. There is a triangular area of whiteness with its apex at the hilum (4) and its base running between the sternum and the diaphragm (5).

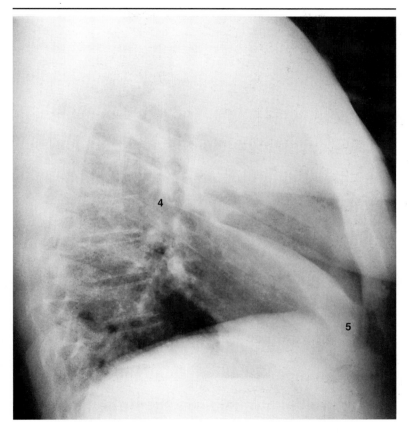

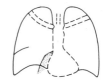

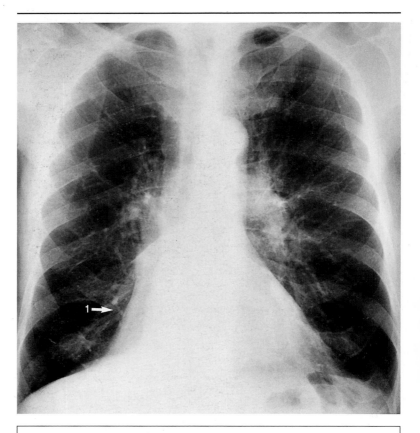

3. ***Right lower lobe collapse.*** *There is a whiteness immediately above the diaphragm (1) causing a loss of its outline. On the lateral film there is a white triangle at the lower posterior part of the lung field (2).*

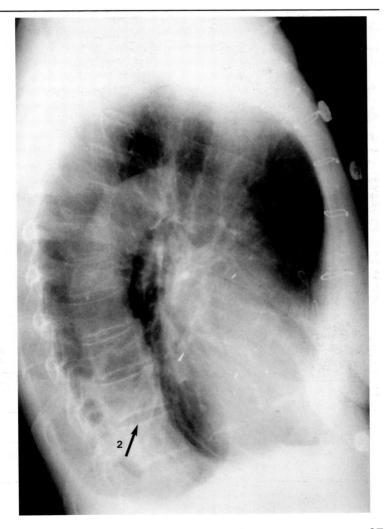

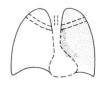

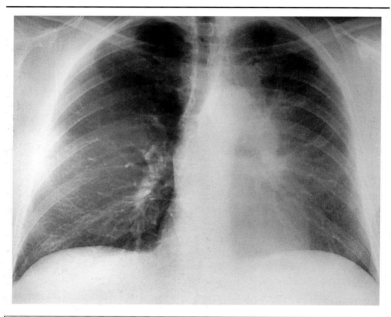

4. **Left upper lobe collapse.** This is difficult to spot. Remember that most of the left upper lobe lies in front of, as opposed to above, the left lower lobe. When it collapses it causes a haze to appear over the whole of the left lung field. On the lateral film an area of whiteness can sometimes be seen at the very top of the lung fields (1).

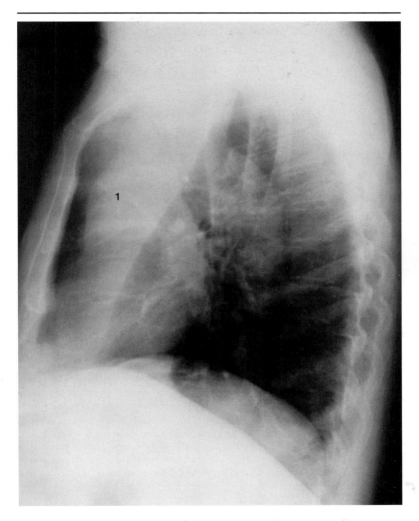

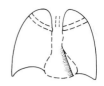

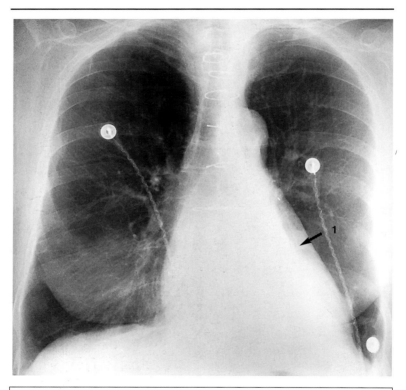

5. **Left lower lobe collapse.** This is easy to miss. The left lower lobe collapses down behind the heart. The left lung field appears much darker than normal and the heart shadow will appear much whiter than normal. If you look carefully you can see a white triangle behind the heart (1). On the lateral film you may see a white triangle at the bottom posterior corner of the lung fields (2) and the vertebral bodies will appear whiter.

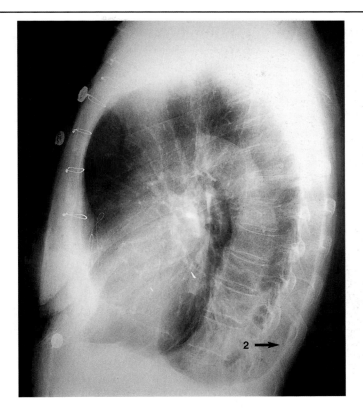

2 →

Volume loss

A more subtle concept is that of volume loss. Many pulmonary processes, including collapse, may cause progressive amounts of volume change during their evolution. Volume loss will decrease the vascularity of the unaffected part of the lung which will look blacker as a result. This is a difficult sign to detect.

Pneumonectomy

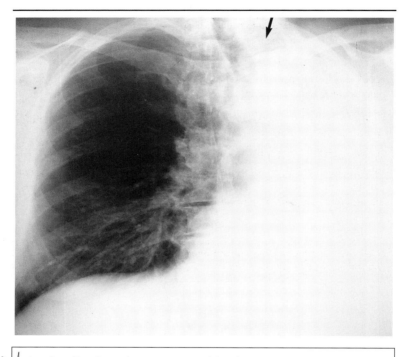

This chest film shows the appearance of the chest in a patient who had a left pneumonectomy five years previously. The left hemithorax is white, the mediastinum shifted to the side of the operation and some of the right lung has 'herniated' to the left side giving a very slightly darker left apex (arrow) compared to the base.

A pneumonectomy is another cause of a white lung. You should know from the history and your examination that the patient has had a pneumonectomy. Look at the X-ray for the following features:

1. Look at the mediastinum. Look first at the trachea which should be shifted to the side of the pneumonectomy. Then look at the heart border. With a pneumonectomy the heart is often so far shifted that its border is no longer visible.

2. Look at the opposite lung field. Since the mediastinum is shifted the contralateral lung is hyperinflated and so appears darker than usual.

3. Look at the side of the whiteness. You should not be able to see the upper border of the diaphragm on the side of the pneumonectomy.

4. Look carefully at the ribs. If the patient has had a pneumonectomy, ribs would either have been cut or removed during the operation. Therefore look for any rib deformity or note the absence of any rib which would help confirm the diagnosis. The most usual rib to be affected is the 5th.

> A very rare cause of a similar appearance is extensive hypoplasia or congenital absence of one lung.

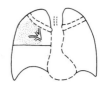

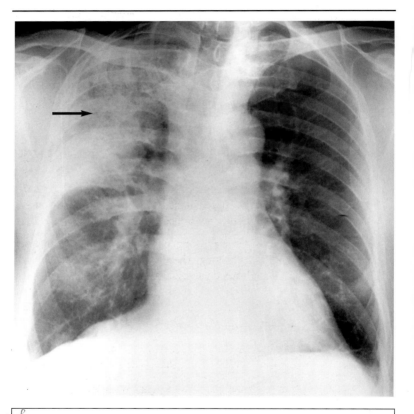

This X-ray demonstrates the typical appearance of consolidation following a lobar pneumonia. An air bronchogram is visible (arrow).

Again you can see an area of white lung. Look first at the nature of the whiteness and its border. If it is uniform with a well-demarcated border you are much more likely to be dealing with an area of collapse or a pleural effusion. If the shadowing is not uniform and the border is not so well demarcated the possibilities are consolidation, fibrosis or some other infiltrative condition. It can be difficult to diagnose consolidation so make your way carefully through the following steps:

1. Remember the clinical history. In the presence of a temperature and signs of infection, consolidation is by far the most likely abnormality.

2. Look at old X-rays. Fibrosis is usually a chronic condition and consolidation much more transient. The presence of a similar abnormality on a previous X-ray should lead you to suspect fibrosis rather than consolidation.

3. Look carefully at the nature of the shadowing. In consolidation the alveolar spaces become filled with fluid making them appear white whereas the airways retain air, making them appear black. If you look closely at an area of consolidation you can often make out the small airways as black against a white background — the so-called air bronchogram.

4. Look at the distribution of the shadowing. Fluid sinks so consolidation gets denser as one moves down the lung. The shadowing in consolidation will often be denser and more clearly demarcated at its lower border.

Consolidation and follow-up X-rays

Any patient with a pneumonia severe enough to ensure admission to an intensive care or high dependency unit needs daily chest X-rays to ensure that their condition is not progressing. For all other patients, daily chest X-rays are not required and a follow-up chest X-ray (at 6 weeks) is only required if the patient has persistent symptoms or signs or is at risk of a lung malignancy.

Pneumocystis carinii pneumonia (PCP)

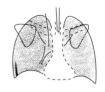

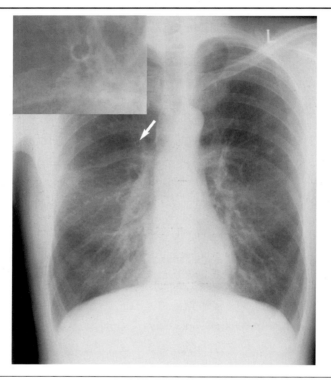

This 24-year-old man was admitted with a 2-year history of increasing shortness of breath. He denied any of the risk factors for HIV and was not receiving treatment with any immunosuppressive drugs. The chest X-ray shows an area of vague white shadowing starting around the hilum and radiating outwards. The enlarged view shows peribronchial cuffing with a thickened bronchial wall (arrowed). The patient went on to have a bronchoalveolar lavage that confirmed the diagnosis of Pneumocystis carinii pneumonia (PCP). He declined an HIV test and discharged himself from the hospital before other investigations had been completed.

Pneumocystis carinii pneumonia (PCP) can be difficult to diagnose on a chest X-ray and in 10% of patients with PCP the chest X-ray is normal. It is something to suspect if a patient presents with shortness of breath and hypoxia which are out of proportion to a relatively normal looking chest X-ray.

If you suspect that the chest X-ray shadowing may be due to PCP then look for the following features:

1. Look at lung volumes. Very early PCP may be suspected when both lungs show reduced volume. Look for an old chest X-ray. If there is one compare the lung volumes to a chest X-ray taken when the patient was well.

2. Look at the area around the hilum. In PCP there is often white shadowing in this region. This can be very vague. You can best appreciate it by looking at the blood vessels as they come away from the hilar. In PCP the blood vessels will appear less well defined than normal.

3. Look for peribronchial cuffing. This is best seen in airways which are seen 'end-on' and is due to fluid seen as increased whiteness within the walls of the airways. This gives the appearance of a white line NOT more than 1.5 mm thick around the airway. It looks like a well-sucked 'Polo' mint.

4. Look for large areas of whiteness extending throughout the lung fields. These may develop as the disease progresses and represent large areas of consolidation. Typically, in PCP the whiteness does not extend to the apices or affect the costophrenic angles.

Asbestos plaques

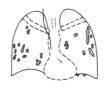

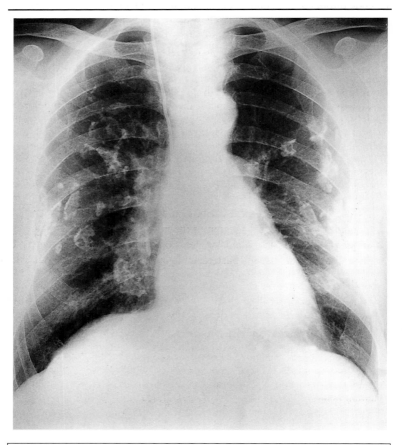

This is the chest film of a 63-year-old man admitted following a gastro-intestinal haemorrhage (hence the central venous pressure line) who had been exposed to asbestos whilst working in a naval dockyard. You can see multiple calcified pleural plaques on the inner chest wall.

Pleural plaques represent areas of pleural thickening caused by asbestos fibres. Isolated pleural thickening is a cause of a localized area of white lung and can be difficult to separate from lung shadows. If you suspect pleural plaques then:

1. Look throughout the lung fields of both lungs. Pleural thickening is easy to identify at the periphery where it appears as a thickened line around the edge of the lung. If you can identify pleural thickening here it makes it more likely that the other areas of whiteness are plaques superimposed over the lung field.

2. Carefully look at the position of the whiteness and compare it to what you know of the anatomy of the lung. If the whiteness follows intrapulmonary structures, for example a lobe of the lung, then it may be originating from the lung itself. If it crosses such structures then it is probably pleural.

3. Compare the distribution of the whiteness to the line of the anterior portion of the ribs. Asbestos plaques are very commonly found running along the line of the anterior portion of the ribs.

4. Look at the distribution of the patches. Pleural plaques are most prevalent in the midzones and axillary region of the midchest. They tend to spare the upper zones and costophrenic angles. Look carefully at both lung fields. Pleural plaques are usually bilateral and you should be wary of making this diagnosis if they are present in only one pleural cavity.

5. Look at the diaphragm. Pleural plaques along the diaphragm are often calcified. If you see streaks of dense white material (calcium) running along the diaphragm it implies that pleural plaques are present.

6. Look at the nature of the whiteness. Pleural plaques are patchy in nature as opposed to a pleural effusion which is more uniform. Plaques are sometimes said to have a map-like appearance due to areas of patchy calcification within them. Look at their edge. This should be well defined as opposed to 'companion shadows' that have poorly defined margins.

7. Look at old X-rays. Pleural plaques are slow growing and are probably visible on previous X-rays.

Mesothelioma

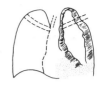

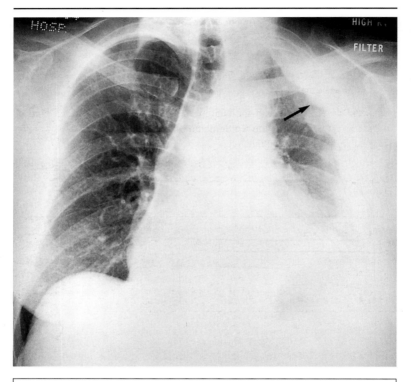

This is the chest X-ray of a 72-year-old man who developed severe left chest pain. It shows the typical features of a mesothelioma with a pleural mass having an irregular, lobulated margin (arrowed) with associated reduction in volume of the hemithorax and crowding of the ribs.

Mesothelioma is a malignant tumour of the pleura. The shadowing it causes will have the characteristics of pleural shadowing and some of the characteristics of malignant shadows. If you suspect the whiteness to be mesothelioma then:

1. Look carefully at the spread of the whiteness and determine whether it follows lung boundaries. If it does not, then the whiteness may be pleural in origin.

2. Look at the edges of the whiteness. If they are lobular in nature then this suggests malignancy.

3. Look at the upper edge of the whiteness. The main differential is a pleural effusion. A pleural effusion would be unlikely if the upper edge was lobular and no meniscus was visible.

4. Compare the volume of the affected side. Loss of volume on the affected side may increase your suspicions of a mesothelioma.

Pleural tumours

Mesothelioma
Secondaries
Pleural sarcoma (very rare)
Benign tumours (rare)

The coin lesion

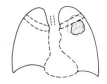

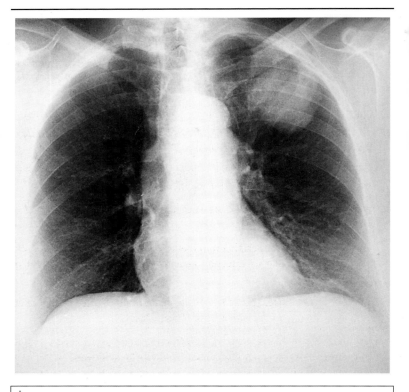

This is the routine preoperative film of a 65-year-old male smoker. There is a mass in the left upper lobe with a well-defined, slightly lobulated border. This is a carcinoma.

The term coin lesion is used to describe a discrete area of whiteness situated within a lung field. It is not necessarily strictly circular. The main worry is that it may represent a carcinoma. Other possibilities are a localized area of consolidation, an abscess or a pleural abnormality. Go through the following steps in assessing the abnormality:

1. Look at the edge of the lesion. A spiculated, irregular or lobulated edge is suggestive of malignancy.

2. Look for areas of calcification. These would be dense white (the same density as bone) and be obviously much denser than the rest of the lesion. Calcification is rare in a malignant lesion and would point you to an alternative diagnosis.

3. Look at the nature of the whiteness. If the lesion is cavitating the centre may be darker than the circumference. Stand back from the X-ray since a cavity is often easier to see from a distance (see p. 11).

4. Look for an air bronchogram. This is a sign of consolidation and so would be a most unusual finding if the lesion was a tumour.

5. Look for other coin lesions. The presence of more than one strongly suggests metastatic disease.

6. Look for abnormalities peripheral to the lesion. A tumour may cause problems distal to it such as infection causing consolidation or an area of collapse.

7. Look carefully at the rest of the X-ray. Malignant tumours may be associated with mediastinal lymphadenopathy or bone metastasis.

8. Look at old films if available. Tumours grow and so if the lesion was present on an earlier film compare its size. Some tumours grow slowly but it is safe to say that if the lesion has not changed over a period of two years or greater it is not malignant.

Causes of single coin lesions

Benign tumour, e.g. hamartoma
Malignant tumour, e.g. bronchial carcinoma, single secondary
Infection, e.g. pneumonia, abscess, tuberculosis, hydatid cyst
Infarction
Rheumatoid nodule

3.9

Cavitating lung lesion

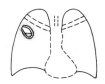

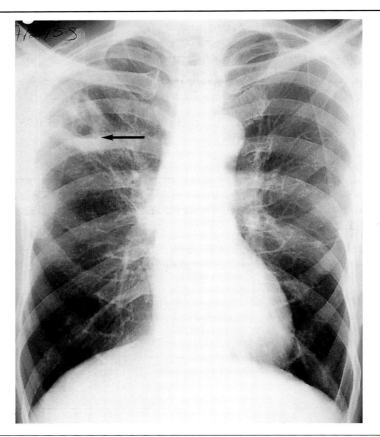

This is the film of a 55-year-old man with a long history of alcohol abuse who developed a chest infection with fever, malaise and cough with foul sputum. There is a whiteness in the right upper zone with the formation of a thick walled cavity which contains an air fluid level (arrow). This is a lung abscess.

Some coin lesions may cavitate and if you have identified a coin lesion it is important to look for the features of cavitation. Therefore:

1. Look at the centre of the lesion and compare it to the periphery. If the centre is darker this points to cavitation.

2. Look for a fluid level. Look for a horizontal line within the lesion. There will be whiteness (fluid) below the line with an area of black (air) above. Fluid levels are common in cavities and their presence should suggest one.

3. Look at the lateral film. Cavities and fluid levels are often easier to see on a lateral, especially when they are posterior or inferior in position.

4. Look at old films. If the lesion is long standing it may be possible to see the cavity developing.

If you diagnose a cavitating lesion:

1. Look at the wall of the cavity. It is often said that cavity walls are thicker (>5 mm) when the lesion is a neoplasm as opposed to an abscess. This rule does not always hold but the thicker the wall the more likely it is that the lesion is neoplastic.

2. Look carefully at the inside of the cavity. If there appears to be a white ball within it this is characteristic of an aspergilloma. This is a rare finding.

Causes of cavitating lung lesions

Abscess
Neoplasm
Cavitation around a pneumonia
Infarct
Rheumatoid nodules (rare)

Left ventricular failure (LVF)

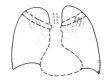

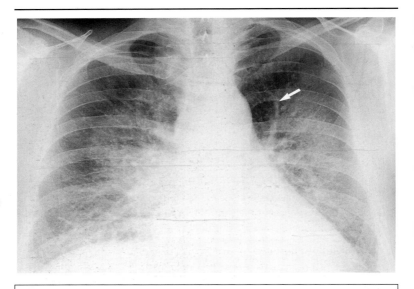

This patient has previously had a coronary bypass graft but now has left ventricular failure. You can see that the heart is enlarged, Kerley B (septal) lines are seen peripherally at each base and the upper lobe blood vessels are dilated and prominent (arrowed). There is a small pleural effusion and this patient's film also shows 'bat's wing' hilar shadows characteristic of pulmonary oedema but a sign that is infrequently seen. The magnified image of the right lower zone (opposite) shows the horizontal septal lines more clearly (arrowed).

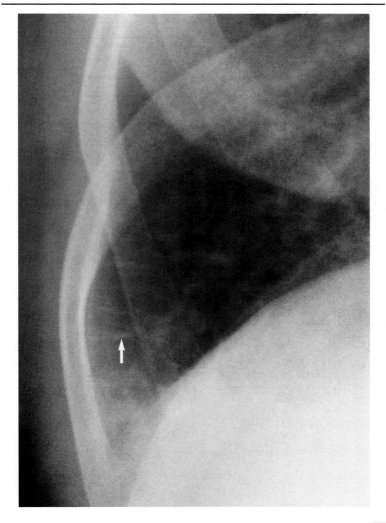

If you suspect heart failure as a cause of a generalized, or localized, area of shadowing then:

1. Compare the size of the upper lobe and lower lobe blood vessels. Take an upper lobe and a lower lobe blood vessel at similar distances from the hilum and compare their widths. The upper should be narrower than the lower. If they are the same size or the upper is wider there is upper lobe blood diversion — the first sign of heart failure. Note that this only applies to an erect film. Upper lobe blood diversion is normal on a supine film and not suggestive of heart failure — a common mistake.

   Upper lobe blood diversion is due to lower zone arteriolar vasoconstriction secondary to alveolar hypoxia.

2. Look at the size of the heart. The presence of left ventricular dilatation is strongly suggestive of heart failure. In a PA film the maximum diameter of the heart should be less than half that of the maximum diameter of the thorax (the cardiothoracic ratio). If it exceeds this then there is a cardiac abnormality such as left ventricular enlargement and this suggests that the associated shadowing is due to LVF (see Chapter 6 for other causes of an enlarged cardiac shadow). You cannot comment on the size of the heart on an AP portable film. Note also that in acute onset LVF you may not get cardiac enlargement.

3. Look for Kerley B lines. These are caused by oedema of the interlobular septa. They are horizontal, non-branching, white lines best seen at the periphery of the lung just above the costophrenic angle.

Severe heart failure

Severe pulmonary oedema gives confluent alveolar shadowing which spreads out from the hilum giving a 'bat's wing' appearance. If this is the cause of generalized shadowing then upper lobe blood diversion and Kerley B lines should be present. Look also at the hilum. In pulmonary oedema it may appear distended and the vessels close to the hilum may be blurred.

Severe heart failure vs non-cardiogenic pulmonary oedema

It can be difficult to distinguish non-cardiogenic pulmonary oedema (acute respiratory distress syndrome) from LVF. In non-cardiogenic pulmonary oedema the heart size is likely to be normal (though see 2 above) and there will not necessarily be sparing of the peripheries. (See also Chapter 3.11)

3.11

Acute respiratory distress syndrome

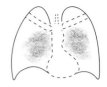

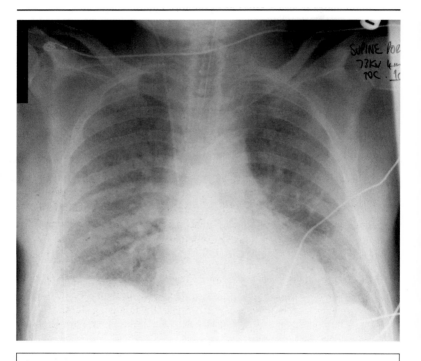

This 36-year-old man presented with severe upper abdominal pain and was diagnosed as having pancreatitis. Whilst in hospital he became acutely short of breath and a chest X-ray was taken which suggested that he had developed the complication of acute respiratory distress syndrome (ARDS). Note that the X-ray shows multiple bilateral white shadows but does not have any signs of left ventricular failure. Unfortunately this man died after 3 weeks of mechanical ventilation on the intensive care unit.

Acute respiratory distress syndrome (ARDS) is defined as respiratory failure in association with a chest X-ray that shows confluent alveolar opacification (whitening) of the lungs that looks like pulmonary oedema.

The other (and far more common) cause of pulmonary oedema is left ventricular failure which is also a cause of respiratory failure. Therefore, if you see a chest X-ray that has bilateral white shadows in the lungs and you suspect ARDS is the cause you need to:

- Check that white shadowing is consistent with a possible diagnosis of ARDS.

- Examine the X-ray carefully to decide whether it is ARDS or left ventricular failure.

- Look for possible clues as to the cause of ARDS.

1. To check that the shadowing is consistent with ARDS look first at its distribution. In ARDS it should be present in both lungs — that is part of the definition. Look also at the nature of the shadowing. It is usually fairly ill-defined which means it is difficult to see a clear edge. It may also have the features of consolidation which means that you may be able to see an air bronchogram within it (p. 44).

2. If the shadowing is consistent with ARDS it suggests that there is a lot of fluid within the lungs. You now need to decide whether this is due to ARDS or left ventricular failure. Look at the X-ray for the following clues:
   a. Look at the heart size. In left ventricular failure you would expect the heart to be big. In ARDS it should be normal size.
   b. Look at the upper lobe blood vessels. In left ventricular failure these will be enlarged (see p. 56) whereas in ARDS they should be the normal size.
   c. Look again at the distribution of the shadowing. In left ventricular failure it tends to be more central whereas in ARDS it is more peripheral.
   d. Look for Kerley B lines (see p. 57). Although these do occur in ARDS they are far more common in left ventricular failure.

e. Look for the presence of a pleural effusion. These may be small so look carefully at the edge of the diaphragm for loss of the normal costophrenic angle. Pleural effusions can occur in ARDS but they are much more common in left ventricular failure.

f. Compare the timing of the X-ray changes to the onset of symptoms. To help you do this look at old chest X-rays if possible. In ARDS the X-ray may not change until 12 hours or more after the development of symptoms. In comparison, in left ventricular failure the chest X-ray often shows abnormalities before symptoms develop.

If you are certain that the chest X-ray suggests a diagnosis of ARDS then look for clues as to the cause. There are many causes of ARDS (see Box on p. 63) and most can only be picked up by history and examination. However, an asymmetrical distribution of the shadowing, i.e. significantly more shadowing in one lung than the other, may point to lung injury as a cause. Chest X-rays taken just before the development of ARDS may show an obvious pneumonia.

Causes of ARDS

ARDS is caused by any insult to either the alveolar or endothelial cells that results in a loss of integrity of the junction between those cells allowing fluid to leak into the alveoli. Causes include:

Aspiration including toxic gas inhalation
Pneumonia
Sepsis
Drowning
Hypersensitivity reactions

Lung trauma
Radiation
Drugs, including drugs of abuse
Fat embolism after trauma
Transfusion reactions

Complications of ARDS

A patient with ARDS should have their chest X-ray repeated daily. Look for signs of disease progression or resolution. Look also for the development of a pneumothorax or lung cysts due to barotrauma caused by the use of positive pressure ventilation in the treatment of ARDS.

Bronchiectasis

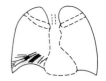

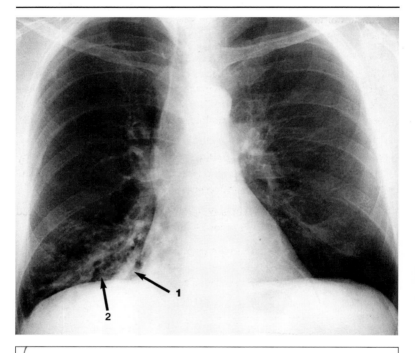

This film shows a localized area of bronchiectasis in the right lower lobe which has resulted from earlier pertussis infection. In the lower lobe you can see typical ring shadows giving a 'bunches of grapes' appearance (1) together with tramline shadowing (2).

Bronchiectasis can be difficult to diagnose on a plain chest X-ray. If you suspect it as a cause of increased shadowing then look for the following features:

1. *Ring shadows*. These look like rings and are any size up to 1 cm in diameter. They can be single but usually occur in groups giving a honeycomb or 'bunches of grapes' appearance. Ring shadows represent diseased bronchi seen end on.

2. *Tramline shadows*. Look for these towards the periphery of the lung. They consist of two thick white parallel lines separated by black. It is common to get parallel lines close to the hilum in the normal chest X-ray but these lines are hairline in nature. True tramline shadowing is thicker and is not necessarily close to the hilum. Tramline shadows are diseased bronchi seen side on.

3. *Tubular shadows*. These are solid thick white shadows up to 8 mm wide. They represent bronchi filled with secretions seen side on. They are not common but their presence suggests bronchiectasis.

4. *Glove finger shadows*. These represent a group of tubular shadows seen head on and look like the fingers of a glove — hence the name!

The presence of any of these features suggests the possibility of bronchiectasis. A normal chest X-ray does not however exclude the diagnosis and CT scanning is the most sensitive diagnostic test available.

Causes of bronchiectasis

Structural, e.g. Kartagener syndrome, obstruction (carcinoma, foreign body)
Infection, e.g. childhood pertussis or measles, tuberculosis, pneumonia
Immune, e.g. hypogammaglobulinaemia, allergic bronchopulmonary aspergillosis
Metabolic, e.g. cystic fibrosis
Idiopathic secondary to stasis

3.13 Fibrosis

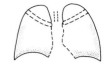

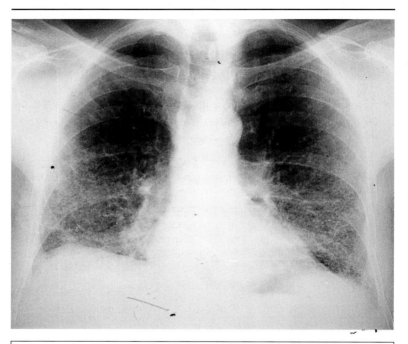

> This patient has cryptogenic fibrosing alveolitis. The X-ray demonstrates many of the features common to fibrotic lung disease. There are fine reticulonodular shadows which extend in a characteristic fashion into the axillary aspect of each hemithorax. The lung volume decreases with progressive fibrosis affecting the lower lobes. The fibrosis may become more prominent leading to the so-called honeycomb appearance consisting of numerous very small ring shadows.

Fibrosis is one of the rarer causes of white lung and you need to differentiate it from consolidation or oedema which are far more common. If you suspect fibrosis:

1. Look at old X-rays if possible. Fibrosis is a fairly chronic process so if present a while ago it is more likely to be fibrosis than consolidation or oedema.

2. Look at the distribution of the shadowing. This may help differentiate fibrosis from oedema since the latter is more likely to be bilateral and basal. Shadowing that is bilateral and basal could be either oedema or fibrosis. Shadowing that is midzone or apical is more likely to be fibrosis.

3. Look at the size of the lungs. Fibrosis may cause shrinkage of the lungs which will not be caused by consolidation or oedema. The presence of small lungs points strongly to fibrosis.

4. Look at the position of the mediastinum. Since fibrosis causes shrinkage of the lungs it will pull the mediastinum to the side of the fibrosis. Shift of the mediastinum towards the shadowing is very suggestive of fibrosis.

5. Look at the nature of the shadowing. Pulmonary fibrosis gives reticular–nodular shadowing which simply means a meshwork of lines that combine to form nodules and ring shadows of about 5 mm in diameter. Sometimes the meshwork is very fine giving a ground-glass appearance, said to look like a thin veil over the lung. Later it gives a more coarse appearance and is said to look like a honeycomb. Compare this appearance to that of pulmonary oedema (p. 56) or consolidation (p. 44) and you will see that it is quite distinctive though in fact it is the other features of fibrosis outlined above that are often the most useful in making the diagnosis.

6. Look at the heart border and diaphragm. Both of these may appear blurred if fibrosis is present.

7. Look at the vascular markings. These become less distinct in areas of fibrosis. This is due to the development of numerous small areas of lung collapse.

Causes of fibrosis

Cryptogenic
External/occupational, e.g. extrinsic allergic alveolitis, asbestosis
Infection, e.g. tuberculosis, psittacosis, aspiration pneumonia
Collagen vascular, e.g. rheumatoid arthritis, SLE
Sarcoid
Iatrogenic, e.g. amiodarone, busulfan, radiotherapy

3.14

Chickenpox pneumonia

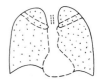

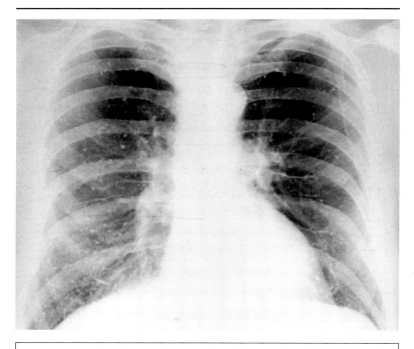

This film of a woman admitted with an acute abdomen shows the typical appearance of old chickenpox pneumonia. You can see numerous bilateral calcified intrapulmonary nodules.

Chickenpox pneumonia in adulthood can cause the development of numerous calcified nodules. To determine whether this is a likely diagnosis:

1. Look at the distribution of the nodules. In chickenpox pneumonia they tend to be lower and midzone.

2. Look at the density of the nodules. They are calcified and so should be very white in appearance.

3. Look at their size. They are usually less than 3 mm in diameter.

4. Look at the number. In chickenpox pneumonia you would expect to see less than 100 nodules. If there are obviously a lot more you should question this as a diagnosis.

Causes of numerous calcified nodules

Infection, e.g. TB, histoplasmosis, chickenpox
Inhalation, e.g. silicosis
Chronic renal failure
Lymphoma following radiotherapy
Chronic pulmonary venous hypertension in mitral stenosis

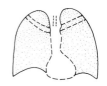

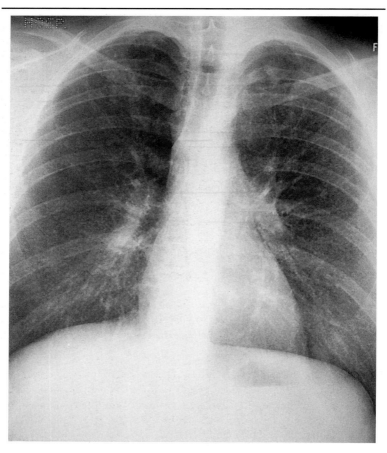

This is the chest film of a 32-year-old man with immunodeficiency. It shows the typical miliary shadowing characteristic of tuberculosis (TB). There is also soft shadowing in the left apex consistent with TB.

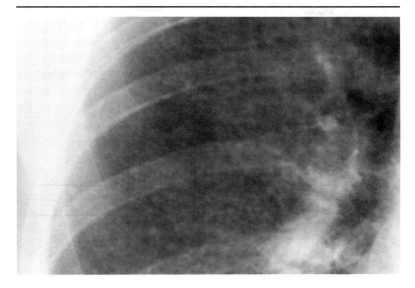

The lungs have a spotted appearance. This may be due to miliary shadowing but sometimes a ground glass appearance can look similar. Also the normal lung can sometimes have a mottled appearance. You first need to distinguish between these.

1. Look at the distribution of the shadowing. Look carefully at the periphery. If the shadowing is present in the periphery it is far more likely to be pathological. Sometimes normal vasculature can mimic interstitial shadowing but this usually occurs only towards the centre of the lung fields.

2. Move close to the X-ray and carefully examine the shadowing. With miliary shadowing the opacities should be discrete.

3. Compare a number of the opacities. If the shadowing is miliary they should be of similar density and size. If the shadowing is ground glass their size and density will vary.

4. Look at the underlying lung field. The normal anatomy of the lung should still be visible with miliary shadowing but may be obscured with a ground glass appearance.

If you feel the shadowing is miliary then look for clues as to its cause. Likely possibilities are miliary TB, sarcoid or malignant miliary metastasis:

1. Look again at the distribution. With miliary TB the opacities are most profuse in the upper zone, with sarcoid they are most profuse in the perihilar and midzones and with miliary metastasis there is a profusion of opacities in the lower zone.

2. Look at the density. High density, very white shadows, are likely to be dust related industrial disease or calcified TB. Less dense changes could be multiple secondaries, sarcoid or any of the other causes of miliary mottling.

3. Look at the rest of the X-ray for signs of other disease processes. Look at the hilum. Unilateral hilar enlargement suggests TB and bilateral hilar enlargement sarcoidosis. Look at the upper part of the mediastinum for thyroid enlargement which could suggest secondaries from a thyroid carcinoma. Look at the apices for subtle cavitating lesions suggestive of TB. Note however that the presence of apical shadowing, although suggestive of TB, is in fact very rare in patients with miliary spread.

Chapter 4

The black lung field

Chronic obstructive pulmonary disease (COPD)

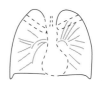

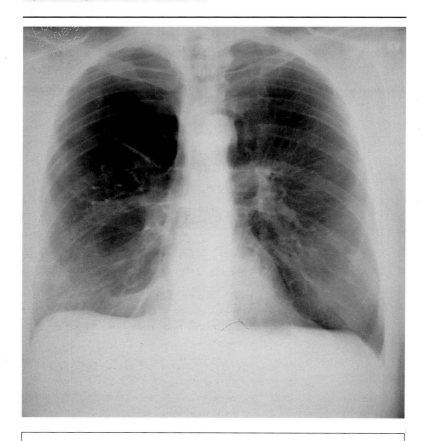

This is the chest film of a patient with chronic obstructive pulmonary disease. Both lungs appear blacker and larger in volume than normal. The hemidiaphragms are flattened and there are bullae in both midzones. Fewer blood vessels are visible peripherally, especially in the upper and middle zones, but the pulmonary arteries are large centrally, consistent with developing secondary pulmonary arterial hypertension.

When trying to decide the cause of *bilateral* black lungs you need to:

1. Check the penetration. Look at the vertebral bodies behind the heart. Remember that in a good quality X-ray the vertebral bodies become harder to see behind the heart shadow. If they are too clearly seen the film is over penetrated making the lungs appear black.

If you are satisfied with the technical quality of the film then the most likely cause is COPD. COPD is associated with large lungs due to air trapping and the development of bullae. You therefore need to:

1. Count the number of ribs you see anteriorly. If the lungs are enlarged you should be able to count more than seven. Be careful however because you can sometimes count more than seven ribs in normal patients.

2. Look at the shape of the diaphragm. In COPD the diaphragms are flat or even scallop shaped instead of concave upwards. This is a more reliable sign of hyperexpansion than rib counting.

3. Look at the shape of the heart. The enlarged thorax of COPD appears on the X-ray to elongate and narrow the heart, elevating the lower border. The heart, instead of sitting on the diaphragm, often appears to 'swing in the wind'. It will also appear small unless there is also an element of cardiac failure in which case it will be normal in size or even large.

4. Look for bullae. These are densely black areas of lung, usually round, surrounded by hairline shadows. Bullae compress the normal lung and distort the surrounding vasculature so to help find them look out for areas of distortion of vascular markings.

5. Look at the distribution of lung markings. The black lungs of COPD are due to decreased lung markings. The lung markings are reduced bilaterally and fan out in straight lines from the hilum, starting off chunky but stopping two-thirds of the way out — peripheral pruning.

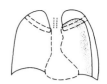

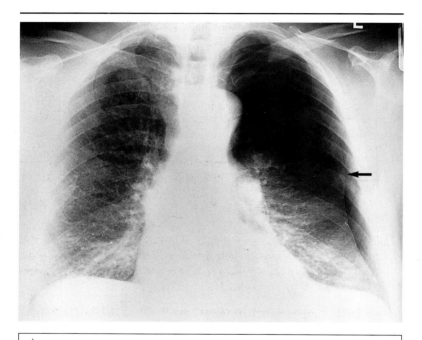

This patient has a left-sided pneumothorax with partial collapse of the left lung. The outer left lung field is black. You can see the lung edge (arrowed).

When you see a *unilateral* black lung you need to:

1. Check the technical quality of the film. A rotated film may make one side less dense than the other.

2. Determine which side is abnormal. This is usually easy since the side with reduced lung markings will be the abnormal side.

You must now decide the cause of the blackness. Lung markings are vascular in nature and it is their absence that makes the lung look black. Vascular shadows will disappear if the lung is replaced by air which will occur with a pneumothorax or bullous or cystic lung disease or if the vessels are deprived of blood as in a pulmonary embolus. Therefore think pneumothorax, bullous/cyst or pulmonary embolism and:

1. Look for a lung edge. In a pneumothorax you will see the edge of lung which is not normally seen. Look carefully at the upper zone where air will accumulate first. Your eye is trained to see horizontal lines better than vertical so it is sometimes easier to detect the lung edge with the X-ray turned on its side.

2. Look at the mediastinum. Obvious mediastinal shift away from the black lung suggests that a tension pneumothorax is developing. This is a medical emergency and you need to urgently reassess the patient (see also p. 78).

3. Look at the rest of the lungs. Bullous disease is less likely if the rest of the lung is normal.

4. Differentiating between a pneumothorax and a bulla can be difficult and often impossible. Look carefully for lung markings. If you see them crossing the area of blackness you are probably looking at a bulla. If you see lung markings peripheral to the blackness it is also probably a bulla.

5. Ask for an expiratory film. In expiration the lung gets smaller and so a pneumothorax will appear bigger.

Causes of a pneumothorax

Spontaneous
Iatrogenic/trauma, e.g. pleural tap, transbronchial biopsy, central venous line insertion, mechanical ventilation
Obstructive lung disease, e.g. asthma, COPD
Infection, e.g. pneumonia, tuberculosis
Cystic fibrosis
Connective tissue disorders, e.g. Marfan's, Ehlers–Danlos

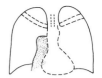

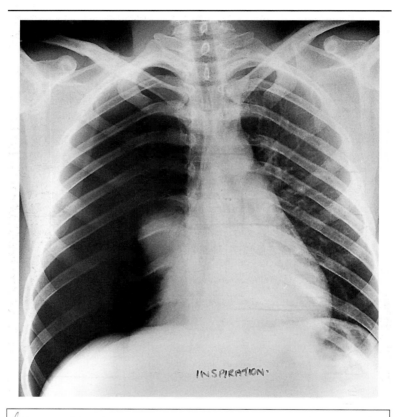

INSPIRATION·

This pair of chest films shows the potentially fatal condition of a tension pneu-mothorax. In inspiration, the right lung is completely collapsed but the media-stinum is central. In expiration, air is trapped in the right hemithorax under positive pressure and the heart and left lung are compressed to the left. Cardiac venous return is obstructed with potentially fatal results if the pleural cavity is not urgently drained.

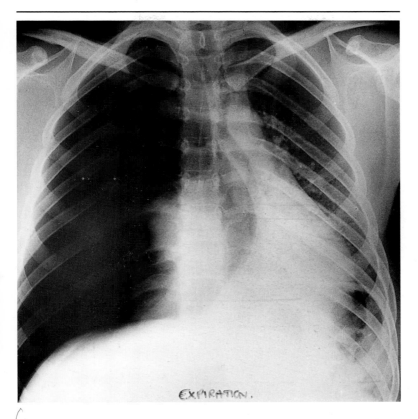

EXPIRATION.

If you suspect a pneumothorax as a cause of a black lung field (see p. 77) you must consider carefully whether it may be under tension since this is a medical emergency. If possible look at an expiratory film and:

1. Look at the size of the blackness. In a tension pneumothorax the black lung is usually very large.

2. Look at the position of the mediastinum. In a tension pneumothorax it will be shifted away from the affected lung.

3. Look at the shape of the mediastinum. Look at the edge on the side of the blackness. If it is concave to the side of the blackness you should suspect a tension pneumothorax.

4. Always remember the patient. A tension pneumothorax can develop at any time and if the patient suddenly becomes distressed the absence of tension on a previous X-ray does not exclude the diagnosis.

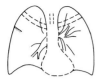

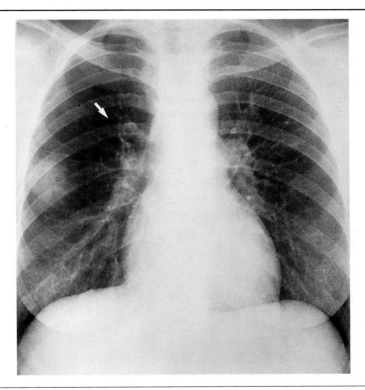

This X-ray is of a patient who has sustained an acute large pulmonary embolus. Look carefully at the right upper zone. Immediately above the horizontal fissure there is an area (arrow) which is blacker than the left side at the same level. This is Westermark's sign of reduced perfusion to that area of lung which indicates that the artery to this area contains a large clot. Note also an area of consolidation below the horizontal fissure — this is a small focus of infarction.

Remember to first check the technical quality of the film. This is especially important since the changes in density caused by a pulmonary embolus are extremely subtle and can easily be produced, or missed, by technical imperfections. This is especially true with portable films.

If you suspect a PE as a cause of a black lung you need to:

1. Check for X-ray signs of COPD or a pneumothorax. You must exclude these since they are far commoner causes of a black lung.

2. Determine whether the area of blackness is confined to segments of the lung and not widespread. An embolus within a pulmonary artery will only affect the segments supplied by that artery and not be a cause of widespread change. It is possible for a massive pulmonary embolus to cause a blackout of a whole lung field but if you suspect this you should stop looking at the X-ray and start treating the patient — they are in imminent danger of death.

3. Look at the rest of the lung. The under perfusion of the affected area will lead to compensatory over perfusion of the rest of the lung and increased density of the vascular shadows. It would be helpful to compare with previous X-rays.

4. Look at the pulmonary artery and the heart shadow. An acute pulmonary embolus will cause dilatation of the pulmonary artery followed by the right ventricle and atrium. The pulmonary artery will therefore increase in size and there may be an increase in the size of the heart shadow.

5. PE is a rare cause of a black lung and usually results in the changes of infarction described below, or in no changes at all. Therefore, unless the patient is very unwell, think again about the other causes of black lung. They are far more likely.

6. Since PEs rarely cause obvious changes on the plain chest X-ray other diagnostic modalities need to be used to make the diagnosis. These will vary from hospital to hospital but include nuclear ventilation/perfusion scanning, spiral CT scanning and magnetic resonance imaging (MRI).

Changes of infarction

Although a PE is a cause of black lung the findings you will usually see following a PE are those due to subsequent infarction of the lung, leading to haemorrhage or lung necrosis. This may cause the following changes on X-ray:

Raised hemidiaphragm
Collapse and linear atelectasis
Pleural effusion
Wedge-shaped shadowing

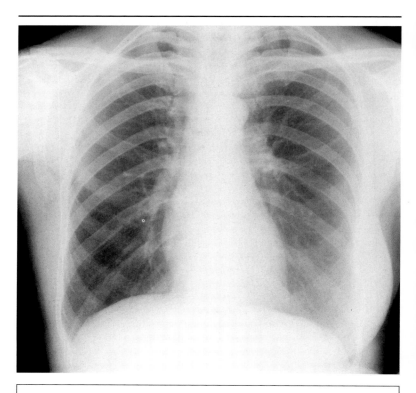

Film of a patient who has undergone a unilateral mastectomy on the left side.

It is important to remember that problems outside the lung can sometimes cause the lung fields to look too black (or too white). This is why it is important to always examine the chest X-ray for soft tissue markings.

A mastectomy will make the underlying lung look too black since there will be less soft tissue overlying the lungs on the affected side, compared to the normal side. Therefore, if one lung looks blacker than the other, look carefully for the breast shadows. There will be an absent breast shadow on the side of the mastectomy.

The breast shadows are easy to detect — the difficulty is remembering to look for them!

Chapter 5

The abnormal hilum

Unilateral hilar enlargement

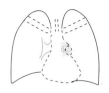

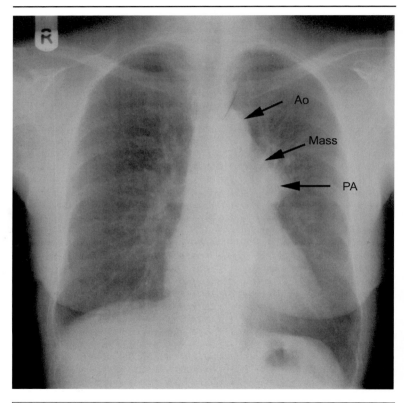

Ao

Mass

PA

This patient has left hilar enlargement. The hilar border is irregular suggesting a malignant cause. The rest of the film is normal. A lateral X-ray is needed and if this confirms that the abnormality is hilar (rather than behind or in front of the hilum) a bronchoscopy is advisable.

Hilar enlargement is difficult! Suspect unilateral hilar enlargement if:

1. One hilum is bigger than the other (obviously — they should be the same size!).

2. One hilum is denser than the other.

3. There is a loss of the normal concave shape — the hila are usually concave in shape. This concavity may disappear and be the first sign of hilar enlargement.

If you suspect unilateral hilar enlargement then:

1. Check the technical quality of the film. A rotated film will make one hilum appear larger than another.

2. Look at the lateral film. An enlarged hilum may look abnormally dense on the lateral and sometimes this is easier to spot than on the PA.

3. Look at the old films. You will be less worried if the X-ray looked the same 15 years ago!

Now you need to decide whether the enlargement is due to enlarged vascular shadows or enlargement of the hilar lymph nodes or whether it is due to a central bronchial carcinoma superimposed over the hilar shadow. These are the likely possibilities.

1. Look at the edge of the hilum. Vascular margins are usually smooth in nature. Lymphadenopathy gives a smooth lobular appearance. Spiculated, irregular or indistinct margins suggest malignancy.

2. Look for the presence of calcium which will appear a very dense white. Its presence suggests lymphadenopathy.

3. Look at the rest of the X-ray. If you suspect hilar enlargement then look carefully at the periphery for lung lesions (tumour, TB), lung infiltration (carcinomatous lymphangitis) or bony lesions (metastasis).

4. Look at the rest of the mediastinum. Malignant hilar enlargement may be associated with superior mediastinal lymphadenopathy.

Hilar enlargement always warrants further investigation.

Causes of hilar lymphadenopathy

Neoplastic, e.g. spread from bronchial carcinoma, primary lymphoma
Infective, e.g. tuberculosis
Sarcoidosis (rarely unilateral)

Causes of hilar vascular enlargement

Pulmonary artery aneurysm
Poststenotic dilatation of the pulmonary artery

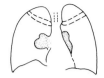

Bilateral hilar enlargement

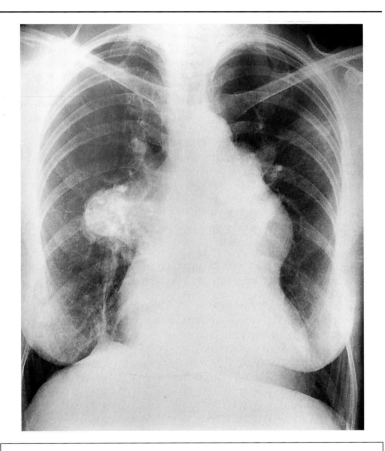

This film shows the typical features of pulmonary hypertension. There is bilateral hilar enlargement and peripheral pruning. Note that the heart is enlarged with right ventricular and right ventricular outflow prominence. This patient has primary pulmonary hypertension.

As with unilateral hilar enlargement, bilateral hilar enlargement can be due to enlargement of pulmonary arteries, veins or lymph nodes. The features that suggest hilar enlargement are described on page 89. In bilateral enlargement they are present on both sides! The commonest causes of bilateral hilar enlargement are pulmonary hypertension and sarcoidosis. You should start off by looking for features of either of these. If you suspect pulmonary hypertension then:

1. Look at the periphery of the X-ray. Pulmonary hypertension is associated with peripheral pruning which means that there is a reduction in peripheral vascularity. The edge of the lung fields are often darker than usual and the central area often whiter.

2. Look at the shape of the hila. It is enlargement of the pulmonary arteries that leads to bilateral hilar enlargement in pulmonary hypertension and so the hilum should take on the shape of enlarged pulmonary arteries. The hila are therefore convex in shape.

3. Look for a cause of pulmonary hypertension. Look for signs of lung disease such as COPD and look carefully at the shape of the heart for signs of left to right shunts or mitral stenosis.

If you suspect sarcoidosis then enlargement of the hilum may be the only finding. However other features are often present and you should look for:

1. *Small nodules*. These are between 1.5 to 3 mm in diameter, are mostly found in the perihilar and midzones, are non-uniform in character, moderately well defined and usually bilateral.

2. *Large nodules*. These are about 1 cm in diameter, have an ill-defined edge and sometimes coalesce to give larger opacities which may contain air bronchograms.

3. *Lines*. The X-ray may demonstrate a network of fine lines emanating from the hilar region.

4. *Honeycombing*. Features of fibrosis may be apparent. Look for these particularly in the upper zones where they are especially common.

Causes of bilateral hilar lymphadenopathy

Sarcoid
Tumours, e.g. lymphoma, bronchial carcinoma, metastatic tumours
Infection, e.g. tuberculosis, recurrent chest infections, AIDS
Berylliosis

Causes of pulmonary hypertension

Obstructive lung disease, e.g. asthma, COPD
Left heart disease, e.g. mitral stenosis, left ventricular failure
Left to right shunts, e.g. ASD, VSD
Recurrent pulmonary emboli
Primary pulmonary hypertension

Chapter 6

The abnormal heart shadow

Atrial septal defect (ASD)

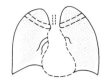

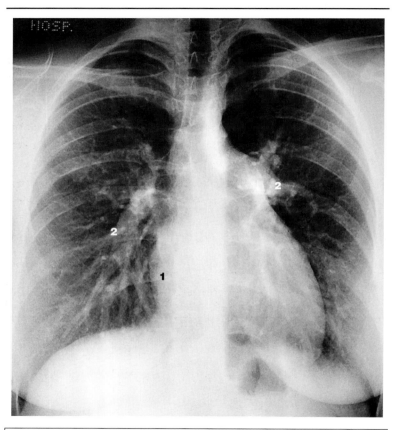

This X-ray shows the typical appearance of an atrial septal defect. The heart is enlarged, the apex is rounded, the right atrium prominent (1) and the pulmonary arteries are dilated (2) due to increased pulmonary blood flow.

Always remember to study the heart and the pulmonary arteries. If the heart is enlarged or pulmonary hypertension is present then one possible cause is an ASD. If you suspect an ASD then look for the following:

1. The heart may be enlarged. Determine the cardiothoracic ratio by measuring the width of the thorax and the width of the heart. If the heart is more than half the diameter of the thorax it is enlarged.

2. Look at the shape of the heart. Look first at the apex which is often rounded due to enlargement of the right ventricle, and is sometimes lifted clear of the diaphragm. Next look at the right heart border. Because the right atrium enlarges, the right heart border looks much fuller than normal.

3. Look at the position of the heart by comparing it to the position of the vertebrae. With an ASD the heart is sometimes shifted to the left and so the right edge of the vertebral column is revealed.

4. Look at the aortic knuckle and arch of the aorta. It is often smaller if an ASD is present since blood is diverted to the right atrium rather than passing through the aorta.

5. Check for the signs of pulmonary hypertension (Chapter 5.2).

An ASD is difficult to distinguish radiologically from other left to right shunts. Echocardiography is the most appropriate means of making a diagnosis.

Mitral stenosis

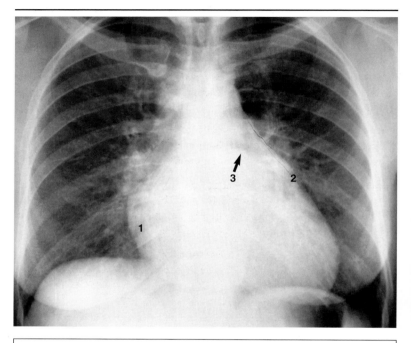

This film is that of a patient who had rheumatic fever when younger. The cardiac contour is abnormal with bulging on the right giving a double right heart border (1), prominence of the left atrial appendage (2) and elevation of the left main bronchus (3), indicating left atrial enlargement. The left atrial pressure is elevated, producing increased pulmonary venous pressure with upper lobe diversion (prominence of the upper lobe veins) and basal septal lines of interstitial oedema (see Chapter 3.10).

Mitral stenosis can cause changes in both the shape and size of the heart. It is a cause of pulmonary oedema. If you suspect mitral stenosis:

1. Look at the left heart border. Look just below the left hilum where the border of the heart is made up of the left atrium (Chapter 2.2). This area is usually concave in shape but in mitral stenosis the left atrium is enlarged causing a loss of this concavity and a straightening of the left heart border. Sometimes atrial enlargement is so great that this part of the heart bulges outwards.

2. Look at the right heart border. Look carefully for a double shadow which you sometimes see in a well penetrated film and is due to left atrial dilatation.

3. Look at the position of the right heart border. Enlargement of the left atrium sometimes causes the right heart border to appear further over to the right than usual.

4. Identify the trachea and follow it down until you see it split into right and left bronchi. This is the carina and the angle between the bronchi should be less than 90°. Measure this angle. If it is more than 90° this is suggestive of left atrial enlargement, a feature of mitral stenosis.

5. Look in the area of the mitral valve for signs of calcification, i.e. flecks of dense white around the valve. This would suggest mitral valve disease but is a rare finding.

Causes of mitral stenosis

Congenital
Rheumatic fever

6.3
Left ventricular
aneurysm (irregular)

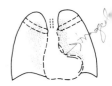

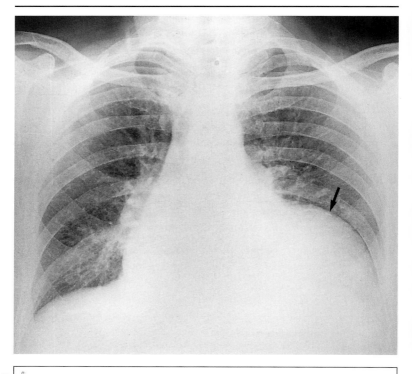

This film is of a 57-year-old man with chest pain and shortness of breath. In addition to the signs of early pulmonary oedema the film also shows a left ventricular aneurysm with an outward and upward bulge of the left ventricular apex. Calcification is also apparent (arrow).

100

A left ventricular aneurysm is a cause of cardiac enlargement on the chest X-ray. It can often cause generalized enlargement of the left ventricle and be indistinguishable from left ventricular dilatation. If you suspect an aneurysm then look for:

1. A bulge in the left ventricle. Follow the left border of the heart. If a part bulges out then this is suggestive of an aneurysm.

2. Look for calcification. If the aneurysm is long standing then it may have become calcified and you will see a rim of calcification along the heart border.

Pericardial calcification

A left ventricular aneurysm is not the only cause of pericardial calcification. Other causes include TB, a calcified infarct and asbestos disease. Often no cause can be found.

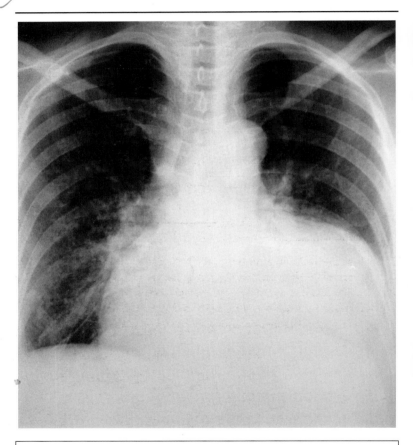

This film shows a pericardial effusion. The heart shadow is enlarged and globular in shape and covers both hila.

Pericardial effusion is a cause of an enlarged heart shadow. If you suspect it then:

1. Confirm that the heart shadow is enlarged. Check that it is a PA film and that the largest diameter of the heart shadow is more than half the largest diameter of the thorax.

2. Look carefully at the shape of the heart shadow. Enlargement due to an effusion is generalized, so if the enlargement appears to be due to a specific chamber enlarging then the cause is unlikely to be an effusion. The heart shadow is globular in shape if an effusion is present, though do not be put off by a bulge that you can sometimes see on the left heart border.

3. Look at the lung fields. If cardiac enlargement is due to LVF then the vascular markings should be increased making the lung fields whiter than usual. In a pericardial effusion the vascular markings are usually normal.

4. Look at previous films. A sudden increase in heart size is suggestive of a pericardial effusion.

5. Look at the hilum. In a pericardial effusion the heart shadow may cover both hila. This will not occur with other forms of cardiac shadow enlargement.

6. Look at the white line on the edge of the right side of the trachea (the paratracheal density). This should be less than 2–3 mm wide on an *erect* chest X-ray. If it is wider, one cause is enlargement of the superior vena cava. This would be consistent with a pericardial effusion (see also p. 102).
See overleaf for 'learning point' on **Causes of pericardial effusions**.

Causes of pericardial effusions

Transudate
   Congestive cardiac failure

Exudate
   Post myocardial infarction
   Infection, e.g. tuberculosis, bacterial
   Neoplastic infiltration
   Collagen vascular, e.g. rheumatoid arthritis, SLE
   Iatrogenic, e.g. postcardiac surgery
   Endocrine — myxoedema

Blood
   Trauma
   Neoplastic infiltration
   Aortic dissection
   Bleeding diathesis, e.g. anticoagulation, leukaemia

The widened mediastinum

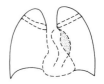

7

The widened mediastinum

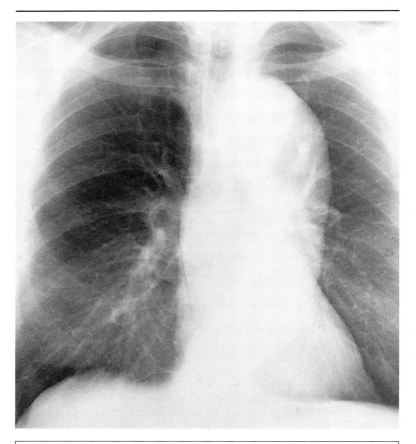

This is the film of a patient with a thoracic aortic aneurysm. The mediastinum is widened throughout its length with increased lateral convexity to the left.

Always look carefully at the mediastinum. If you think that it is widened then relate this finding to the clinical history. If you suspect an acute aortic aneurysm then you must follow up your suspicions as quickly as possible with a CT, echocardiogram or MRI. If an acute event is unlikely you need not move so fast. Try to find some old films and see if the mediastinum has got larger.

Important causes of a widened mediastinum are thyroid enlargement, enlargement of mediastinal lymph nodes, aortic dilatation, dilatation of the oesophagus or thymic tumours. In deciding a likely cause go through the following process:

1. Check the rotation of the film. A badly rotated film can make the mediastinum appear widened.

2. Decide whether the enlargement is at the top, middle or bottom of the mediastinum. If at the top it is likely to be thyroid, thymus or innominate artery. If in the middle or bottom of the mediastinum it could be lymphadenopathy, aortic widening, dilatation of the oesophagus or a hiatal hernia.

3. If the shadowing is at the top then look at the position of the trachea. An enlarged thyroid will displace or narrow the trachea. This will not happen with a tortuous innominate artery — a common finding in the elderly.

4. Look at the right side of the trachea. The white edge of the trachea should be less than 2–3 mm wide on an *erect* film. An increase in its width suggests either an enlarged superior vena cava or a paratracheal mass. This rule does not apply to supine films.

5. If you suspect an enlarged thyroid then look at the outline of the shadow. A thyroid has a well defined outline that tends to become less clear as one moves up to the neck.

6. If you suspect widening of the aorta then try and follow its outline, remembering that the root of the aorta is not visible. You may be able to detect a continuous edge which widens to form the edge of the enlarged mediastinum. This would suggest that the widening is due to dilatation of the aorta.

7. Look for calcification in the wall of the aorta. If you can detect a line of calcium then follow it. If it leads into an area of aortic contour then this strongly suggests an aortic aneurysm. If the line of calcium is separated from the edge of the aortic shadow this strongly suggests a dissection.

8. A widened, aneurysmal, aorta can sometimes be difficult to distinguish from the more common unfolded aorta. If you can follow both edges of the aorta and detect a widening this suggests an aneurysm. Obtain a lateral film. If the edges of the aorta are parallel then you are probably looking at an unfolded aorta on the PA.

> If you suspect an aortic dissection and only have a supine film then the following rule is of use: measure the width of the mediastinum at its point of maximum convexity and compare it to the width of the chest at that point. If it is more than 30% of the width of the chest then the mediastinum is enlarged and aortic dilatation must be suspected.

Chapter 8

Abnormal ribs

Metastatic deposits

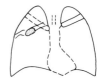

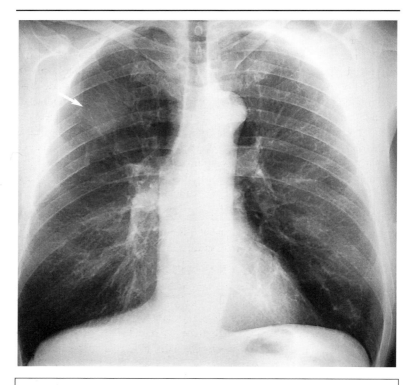

Look carefully at this film. The lungs are overexpanded. Note the destruction of the posterior part of the cortex and the medulla of the right fifth rib with an associated ill-defined soft tissue mass (arrow). This is a <u>lytic metastasis</u>.

Your examination of the chest X-ray is not completed until you have looked carefully at the ribs. They should be of a uniform density with smoothish, unbroken edges. The main abnormalities to look for are old and new fractures and metastases.

1. *New fractures.* Look along the edges of each rib. A new fracture will be seen as a break in the edge. Once you have spotted a fracture look for more information. Look at the position of the fracture. A fracture of any of the first three ribs is unusual and implies tremendous force. Look for other fractures. A line of fractures suggests a traumatic injury whereas fractures scattered throughout the ribs may suggest repeated injury (as in an alcoholic) or underlying bony weakness (as in malignant disease). Look at the density of the ribs and compare them in your mind to other X-rays you have seen. If the ribs are less white than usual this suggests underlying decrease in bone density. Finally look for the complications of rib fractures — surgical emphysema, pneumothorax and haemothorax. Remember also that damage to the lower three ribs may result in hepatic, splenic or renal injury.

2. *Old fractures.* Again look along the edges. The callus formation that follows a fracture will cause the rib to expand at this point. You need to look carefully — sometimes callus formation can simulate a lung mass.

3. *Metastases.* These look like dark holes in the ribs. Scan the ribs carefully for evidence of metastasis. Secondaries start in the medulla and spread outwards with very little reaction around them so you are literally just looking for a dark hole. Sometimes the underlying lung markings create the impression of a metastasis in the overlying rib. If you spot a dark hole in the rib then look carefully at its edge. Compare the outline to the underlying lung markings. If they overlap then the metastasis may be deceptive.

4. Look carefully at the other bones which may contain similar pathology.

Diagnosis of rib fractures

Rib fractures can be missed on a chest X-ray. The diagnosis is therefore clinical and the chest X-ray is usually performed to look for potential complications.

Chapter 9

Abnormal soft tissues

Surgical emphysema

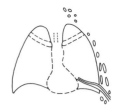

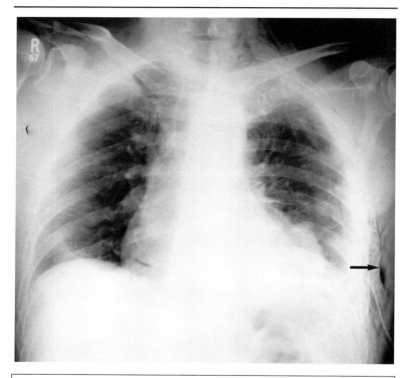

This is the chest film of a 59-year-old man who has had a left chest drain inserted for the relief of a pneumothorax. Although the lung remains fully inflated, air has tracked out from the pleural cavity along the drain and has collected in the subcutaneous tissue (arrow) from the neck to the waist.

At first sight surgical emphysema gives a very messy appearance which is sometimes confined to the obvious soft tissue areas but may spread over the whole X-ray. If you suspect surgical emphysema look for the following characteristics:

1. In mild cases look for lozenge-shaped areas of blackness which represent pockets of air in the soft tissue. These areas will all lie in the same plane which will follow the plane of the soft tissue structures.

2. In severe cases the orientation of the planes is lost. Instead look for alternating dark and white lines which appear not to be confined to single structures and cross part or all of the film.

Causes of surgical emphysema

Trauma
Iatrogenic, e.g. surgery, chest drain insertion
Obstructive lung disease, e.g. asthma
Oesophageal injury
Gas gangrene

Chapter 10

The hidden abnormality

Pancoast's tumour

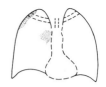

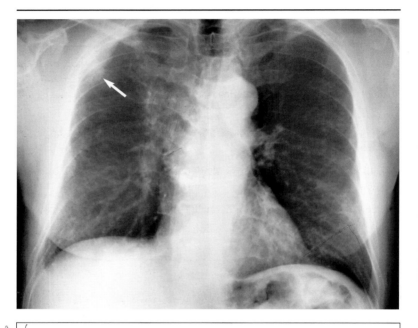

Superficial inspection may suggest that there is little wrong with this film. Note however the increased whiteness at the right apex (arrow) and the haziness at the upper part of the right hilum. Closer inspection reveals that the apical mass is destroying the 1st and 2nd ribs. This is a Pancoast's tumour.

A number of abnormalities can be easily missed. Before dismissing an X-ray as normal:

1. Look carefully at the apices of both lungs. This is a common site for lung pathology, for example a Pancoast's tumour or chronic fibrosis. Lesions here can be easily missed because the apex of the lung is hidden by ribs and clavicles.

2. Look carefully at the heart shadow. Lesions behind the heart are often missed because they are obscured by the whiteness of the heart. Look carefully for any parts of the heart shadow that look whiter than the rest. Look also for the triangular shadow of left lower lobe collapse.

3. Look carefully at the mediastinum. Changes in the shape of the mediastinum can be very subtle.

4. Look at the hilum. Changes in the shape or density of the hilum can be easily missed.

5. Obtain a lateral film — some abnormalities are more obvious on the lateral. Read the radiologist's report!

✓**Exams and the normal X-ray**

Spotting the abnormality in an apparently normal chest X-ray is a common question in postgraduate exams. If confronted with such an X-ray, then think of the following possibilities:

Apical shadowing
Left lower lobe collapse
Hiatus hernia (fluid level behind the heart)
Dextrocardia (with the X-ray shown the wrong way around)
Mastectomy
Air under the diaphragm

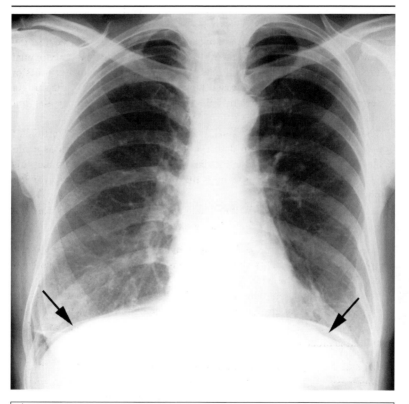

This is the X-ray of a 72-year-old man who presented to casualty acutely unwell with abdominal and chest pain. Examination revealed a silent abdomen. Note the areas of blackness immediately under the diaphragms (arrowed). This represents <u>air collecting under the diaphragm</u> and confirms the clinical suspicion of <u>abdominal perforation</u>. Subsequent history-taking revealed that he had a 2-year history of recurrent upper abdominal pain and on surgery a perforated gastric ulcer was found.

Finish your examination of the chest X-ray by looking at the area under the diaphragm. The area immediately under the diaphragm will usually be white since the upper part of the abdomen contains the dense structures of the liver and spleen. Because of this you can usually only make out the upper surface of the diaphragm You may see a darker round area under the left hemidiaphragm. This is the air bubble within the stomach.

One of the main reasons for looking under the diaphragm is to detect the presence of free air. This is an important sign since it indicates intra-abdominal perforation. Other intra-abdominal pathologies you might see include areas of calcification (small areas of increased whiteness) under the right diaphragm corresponding to gallstones, and dilated loops of bowel under the left diaphragm.

The chest X-ray is a very sensitive investigation for the detection of free abdominal air since it can detect as little as 10 ml. It appears as a rim of blackness immediately under the diaphragm and you will recognize this since it may enable you to see both the upper and lower surface of the diaphragm.

It is sometimes difficult to differentiate air under the diaphragm from the normal stomach bubble. If in doubt then look at the following:

1. Look at the thickness of the diaphragm, that is the line between the blacker area below and the lungs above. If there is free air immediately below the diaphragm, then the white line between the air and the chest will appear very thin since it will consist of the diaphragm only. If the air is in the stomach then the white line created will consist of both stomach, lining and diaphragm and appear thicker. In general, if the line is less than 5 mm, then free air is probably present.

2. Look at the length of the air bubble, that is the distance from its medial to lateral aspect. If it is longer than half the length of the hemidiaphragm it is likely to be free air, since air within the stomach is restricted by the anatomy of the stomach.

3. Look at both hemidiaphragms. If air is present below the right and left hemidiaphragms, it is likely to be free air in the abdomen.

4. If you are still in doubt order a decubitus film. This is taken with the patient lying on their side. Free air will rise away from the diaphragm to the uppermost aspect of the abdomen whereas air within the stomach will remain in the same position. Remember that it takes over 10 minutes for these changes to occur so the patient needs to be on their side for 10 minutes before the X-ray is taken.

Index

Topic headings are shown in **bold** type. Page references in *italics* indicate illustrations.